Junkyards, Gearheads, and Rust

Junkyards, Gearheads, and Rust

Salvaging the Automotive Past

David N. Lucsko

JOHNS HOPKINS UNIVERSITY PRESS   BALTIMORE

Johns Hopkins University Press
2715 North Charles Street
Baltimore, Maryland 21218-4363
www.press.jhu.edu

Library of Congress Cataloging-in-Publication Data

Lucsko, David N., 1976–
    Junkyards, gearheads, and rust : salvaging the automotive past / David N. Lucsko.
        pages   cm
    Includes bibliographical references and index.
    ISBN 978-1-4214-1942-8 (hardcover : alk. paper)—ISBN 1-4214-1942-4 (hardcover :
alk. paper)—ISBN 978-1-4214-1943-5 (electronic)—ISBN 1-4214-1943-2 (electronic)
1. Automobile graveyards—United States. 2. Automobiles—Parts—United States.
3. Automobile wrecking and used parts industry—United States—History. I. Title.
    TD795.4.L83 2016
    338.4′762920973—dc23   2015026986

A catalog record for this book is available from the British Library.

*Special discounts are available for bulk purchases of this book. For more information, please
contact Special Sales at 410-516-6936 or specialsales@press.jhu.edu.*

Johns Hopkins University Press uses environmentally friendly book materials, including recycled text paper that is composed of at least 30 percent post-consumer waste,
whenever possible.

For Walter, and for Bob

Contents

Preface

When I was four, I wanted one toy more than any other: Clyde's Car Crusher, a set that let you form small cars out of tinfoil using molds and then smash them into little cubes with a miniature junkyard crusher. I never got a Clyde's set of my own, but when I was five, I got something even better, at least to my mind. That year my family moved from one borough in the Pittsburgh area to another, and from my bedroom window on the front of our new house I could clearly see, over the crest of a hill on the horizon, the cranes and piles of cars at a salvage yard a couple of miles away. Eighteen months later we moved again, this time to a more idyllic town, south of Atlanta. To my disappointment, all that could be seen on the horizon from my new window was an endless line of southern pines. But to my delight, anytime we drove to the mall in the next county, or to downtown Atlanta itself, I had plenty of opportunities to see derelict cars rusting away in fields or lined up in roadside salvage yards.

I'm not quite sure why I found these sights so fascinating when I was a kid. What I do know is that when I was a teenager, the salvage yards scattered across the southern metro area became vitally important to me as sources of odds and ends for my daily driver. Over the years I bought emblems, fenders, shift knobs, dashboard switches, and several speedometers from the local yards. Sometimes I also visited them just to browse around, poking and prodding at the wrecks as if they were department-store goods and I a window shopper. On more than one occasion, I missed my chance to rescue a vintage car, too; I'll never forget the clean, rust-free, white-on-white 1967 Westfalia camper that was inexplicably moldering away in a yard on Highway 85 or the guilt I felt when I realized that I was in no position to save it from its fate. Yes, once upon a time, I was a junkyard junkie.

In the years since, my trips to salvage yards have grown fewer and farther between. This is because the parts I tend to need for my older cars simply don't exist at most of them anymore; the wrecks that would have had them once upon a time have now aged out of the system, having grown too old to be kept around profitably. But salvage yards have never been far from my mind. While working on my dissertation and my first book, I encountered numerous references in *Hot Rod*, *Popular Hot Rodding*, and other periodicals to the use of parts from salvage yards in everything from street rods to all-out restoration projects. As a result, the thought occurred to me that an examination of salvage yards and the practices of reuse and repurposing common among gearheads might be a good way to follow up on my research on the culture and technology of the high-performance hobby. So I started collecting articles and references on the side. Shortly after my first book went to press, I used what I had gathered to draft a conference paper about clunker programs and zoning rules. Six years then flashed by in a blur, and now I find myself sitting in my office on a lovely spring afternoon, drafting the acknowledgments to *Junkyards, Gearheads, and Rust*.

Were it not for the generous assistance of a number of individuals and institutions, I never would have been able to see this project through. First and foremost, I thank my colleagues in the History Department at Auburn University, especially Bill Trimble, who graciously read and commented on several drafts of chapter 6; Morris Bian, who shifted my scheduled term of leave so that I could finish the final draft of the full manuscript on time; those who attended the faculty seminar in November of 2011, where I presented an early draft of chapter 6; the students who took my graduate seminar on waste and recycling in the spring of 2012; and last but certainly not least, all of the faculty members and graduate students who gather at Halftime each week. For their valuable feedback and suggestions, I also thank those who attended the History, Technology, and Society colloquium at Georgia Tech in November of 2011; the Research Seminar at the Hagley Museum and Library in February of 2013; the meetings in Pittsburgh (2009), Cleveland (2011), and Portland (2013) of the Society for the History of Technology (SHOT); and the Twenty-Fifth Anniversary Symposium of the Doctoral Program in Science, Technology, and Society at MIT in November of 2013. Any errors of fact or interpretation are mine alone, of course.

An abbreviated version of chapter 6 and portions of chapters 2 and 5 appeared in *Technology and Culture* as "Of Clunkers and Camaros: Accelerated Vehicle Retirement Programs and the Automobile Enthusiast, 1990–2009" (April 2014): 390–428. I thank Johns Hopkins University Press for allowing me to reproduce that material here. Portions of chapter 2 appeared in "American Motor Sport: The Checkered Literature on the Checkered Flag," a historiographic essay in *A Companion to American Sport History*, ed. Steven Riess (Wiley Blackwell, 2014), 313–33. I thank the editors at Wiley Blackwell for their permission to reproduce those portions here.

I also thank the anonymous referees at *Technology and Culture* and Johns Hopkins University Press for their feedback and encouragement. For supporting this project from day one and for seeing it through to the end, I thank Bob Brugger, my editor at Johns Hopkins University Press. I also thank Kathryn Marguy and the rest of the staff at the press for everything they've done to bring this volume into print. For their tireless assistance, I thank the librarians at Auburn University; at the University of California, Irvine; at the University of California, Riverside; and The Henry Ford.

For generously allowing me to reproduce their copyrighted images, I thank G. Bugg Jr. of the National Street Rod Association; Eric Kaminsky of Amos Press, Inc.; Michael McNessor of *Hemmings*; Lyle R. Rolfe; Phil Skinner; and George Trosley. For taking the time to either locate archived and unarchived resources, sit for formal interviews, or both, I thank the Specialty Equipment Market Association's (SEMA's) Chris Kersting, Peter MacGillivray, Steve McDonald, and Stuart Gosswein; Alexa Barron of the California Air Resources Board; Rob Kinnan, David Freiburger, and Mike Finnegan of *Hot Rod*; Phil Skinner; and Kem Robertson.

For financial assistance while researching this project, I thank both the College of Liberal Arts and the Department of History here at Auburn. For helping me in so many ways as I juggled this book project, my teaching load, and my work as SHOT secretary, I thank Bruce Seely, Francesca Bray, Bernie Carlson, and Jane Carlson. For countless conversations about all things automotive, and for being such good friends since I first came to Auburn, I thank Dave Burke, Jim Hoogerwerf, Alan Meyer, and Bill Trimble. I also thank Roe Smith, my graduate advisor, for his support over the last decade and a half, as well as Bob Post and Joe Schultz, who patiently listened as I struggled early on to explain why I thought junkyards were a good topic, and then encouraged me to follow through. I thank John Staudenmaier, too, for giving me a shot at *Technology and Culture* a decade ago, and for being such a good friend and mentor ever since. I thank my mother, Susan Lucsko, for putting up with me and all my cars through the years; Mr. Vimbainjer and Korylya deserve a pat on the back as well. I also want to give a heartfelt nod to Danny Lucsko, my father, Scrabble partner, fellow fan of old-school garage rock, and steadfast friend, who succumbed to cancer less than one month after his sixty-second birthday in 2012.

I dedicate this book to two longtime friends: my air-cooled motoring mentor, Walter Donila, who once found automotive treasure underneath a coffin in a salvage yard, and my uncle, Bob Deull, whose barn-stored stacks of crusty magazines helped kick-start this project.

Junkyards, Gearheads, and Rust

Introduction

Slightly more than one hundred years have now passed since the first mass-produced Model Ts rolled off the line at Highland Park. Not coincidentally, slightly more than one hundred years have also passed since salvage yards specializing in automobiles first began to appear. Occupying a narrow niche between the emerging used-car business and the well-established scrap-metal trade, these early salvage operations were straightforward, if unsightly. "We tear 'em up and sell the pieces," as one salvage-yard operator explained, dismantling wrecked and worn out cars both for serviceable parts and systems, to be sold in the growing retail used-parts trade, and for functionally useless odds and ends, to be sold wholesale as scrap. The yards that performed these tasks were fewer and much farther between than the gasoline stations, repair shops, and other businesses rightly credited with enabling automobile use to become widespread in the early twentieth century.[1] But they were no less vital. By the middle of the 1920s, as both the new- and used-car markets became saturated, the number of older models retired from service each year quickly reached into the millions.[2] By absorbing these unwanted relics, this first generation of salvage yards made room for countless brand-new Chevys, Dodges, and Fords, while also helping a good number of used cars stay on the road. In so doing, they played an indispensable role in the maintenance and further development of American automobility.

But it was a thankless task. Like other vital facilities associated with refuse and waste, including scrap-metal yards, sewage treatment plants, garbage transfer stations, incinerators, and landfills, automotive salvage yards have long epitomized the cutesy euphemism "LULU." Short for "locally unwanted land use," a LULU is by definition difficult for county, town, or city officials to fit into their zoning and land-use matrixes. This is often due to the objections of federal and state environmental authorities, but an equally common stumbling block is "NIMBYism": organized resistance— "not in my backyard!"—among those concerned about a given LULU's impact on everything from property values to the general aesthetics and quality of life in their neighborhoods.[3] Although the terminology is relatively recent, neither LULUs nor the NIMBY sentiments confronting them are new. From the ancient habit of throwing one's garbage over the nearest wall to the equally ancient (though in more recent times far more damaging) practice of sending industrial pollution downstream, we have all long searched for what Joel Tarr calls "the ultimate sink," an ideal place to dump our refuse where it will forever remain out of sight and out of mind.[4]

Such has always been the case with out-of-service automobiles. The first to be retired, in the late 1890s and very early 1900s, were either tucked away in the barns and carriage houses of their elite owners or were sold for dismantling at the hands of scrap-metal dealers. This ad hoc disposal system worked well enough when the number of pioneering motorists was relatively low. But during the decade of the 1900s, automobile ownership and usage quickly broadened.[5] As this happened, the demand for inexpensive repair parts rose swiftly, and so did the need for another "sink"— another means of disposing of the growing number of wrecked, hopelessly obsolete, or otherwise unwanted cars. At least in theory, automotive salvage yards were uniquely positioned to address both of these pressing needs. But in practice, these yards proved much better at supplying used parts than at the arguably more important task of keeping their burgeoning inventories out of sight and out of mind. Whether located in a city (common among the earliest yards) or on its outskirts (common by the 1920s), the typical salvage yard was less a place where old wrecks disappeared than where they *gathered*, often in plain sight. Moreover, salvage work itself could be visually, aurally, and olfactorily offensive, not to mention dangerous: the frequent showers of sparks, the ceaseless clamor of heavy tools, the routine bonfires of picked-clean shells, the pervasive smell of gasoline and oil. Such would be the case for nearly fifty years, until a series of legal and technological breakthroughs in the 1960s and 1970s meaningfully altered their day-to-day operations for the better.[6]

But by then the die was cast, at least as far as the sector's public reputation was concerned. Amid swirling allegations of environmental damage, property-value erosion, and ties to organized crime, the salvage business of the 1980s, 1990s, and the early twenty-first century was both feared and loathed—and misunderstood. Consider its depiction in contemporary popular culture. Not for nothing does Quentin Tarantino's masterpiece *Pulp Fiction* nonchalantly feature an underworld body dump in "Monster Joe's" salvage yard. Not for nothing, either, does *Breaking Bad*'s newbie meth cook Walter White, much to the amusement of veteran criminals Jesse Pinkman and Tuco Salamanca, assume that wholesale drug deals ought to happen in seedy places like salvage yards. Indeed, not for nothing do most of those who mention them in movies, television programs, popular music, and literature call them "junkyards" instead of "salvage yards," "automotive dismantlers," or any of a number of other value-neutral alternatives.[7] For what matters isn't their vital economic and environmental functions but that they are unsightly and unseemly. This unwholesome reputation, coupled with the inescapable realities of a business that is in fact dirty and dangerous, helps explain the junkyard's quintessential LULU status. We don't mind one on the silver screen or on the written page. But in real life most of us do everything we can to avoid encountering one. We also bitterly complain whenever anything resembling one turns up in our communities. It's the sort of story on which local news spreads thrive: "Neighbors Upset over Junkyard," "Student Goes Head-On after Junk Cars," "Darby Twp. Tires of 'Junkyard' on Hook Road," "City Clears 'Junkyard' Eyesore," "Block Resembles Junkyard."[8]

The salvage industry itself puts forth a different narrative. In the pages of *Waste Trade Journal* and later *Scrap, Recycling Today, Automotive Industries*, and other trade periodicals, those in the business have long held that salvage yards, scrap yards, and other recycling operations constitute a vital and by-and-large responsibly managed sector of the economy. They're right, of course, but only on occasion does their counternarrative break through to a wider audience. When it does, it generally takes the form of a major newspaper or broad-based magazine running a detailed piece explaining, often with an open sense of surprise, that salvage yards are useful sources of inexpensive parts, or that scrapped cars are eventually melted down to make new cars, or that most salvage and scrap yards actually aren't environmental scofflaws. These stories tend to look and feel remark-ably similar, even when published years or even decades apart. It seems that every few years yet another journalist, having noticed that media treatments of the sector are overwhelmingly negative, proceeds to interview a few key players and unearth essentially the same story: junkyards—ahem, *salvage yards*—really aren't all that bad.[9] And for many of their read-ers over the last half century, an era steeped in antijunkyard coverage, this may well have seemed like news.

A third narrative thrives just beneath the surface. It too runs counter to the pervasive vilification of the salvage business, but it also does not mesh particularly well with the industry's protective discourse nor with the occasional thumbs-up from the likes of the *New York Times*. It seeks to make the most of salvaged cars and parts, that is, but it does not necessar-ily extol the broader economic benefits of a robust salvage sector. Neither does it consistently praise the industry's efficiency improvements, nor its ever more environmentally sound practices, and it certainly does not try to justify the business as a "necessary evil." Instead, this third view, widespread among gearheads[10] since at least the interwar years, actually celebrates what Tom McCarthy calls the "automotive afterlife"—junked cars, junkyards, and everything that they entail.[11]

In the 1990s, for example, well aware that many of his customers en-joyed browsing among the rust and weeds of his thirty-acre yard in Ohio, a second-generation salvage man, John Schulte, "facetiously hung a 'Play-ground' sign" at his gates.[12] At about the same time, Tony Attard, a body-shop owner in suburban Detroit, also operated a five-acre, vintage-specialty salvage yard on the side as his own "after hours playground."[13] Likewise, many of those who visited Bob's Auto Wrecking and Recovery in northern Ohio "weren't in any big hurry" and typically "wouldn't be looking for just one kind of car." Instead, the yard's owners explained in the mid-2000s, "they just enjoy being around the cars—like us."[14] More broadly, enthusi-ast periodicals of every stripe have long brimmed with playful coverage of parts- and parts-car-hunting excursions: "junkyard jamborees," "salvage-yard treasure hunts," and guides to "wreck-connaissance," the "salvage-yard maze," and the hunt for "junkyard jewels."[15] Others have waxed enthusias-tic about rummaging through treasure troves of "junque" at giant swap meets like the one hosted by the Antique Automobile Club of America each fall in Hershey, Pennsylvania. They have also raved in print about countless custom and restoration projects that began as hopeless junkyard hulks, about the creative reuse or repurposing of junkyard-sourced components in a variety of automotive projects, and about the visual and olfactory joys of wandering through a field of wrecks. Some enthusiasts have adorned their office or garage walls with photo calendars and posters featuring rusting hulks in open fields and other idyllic settings or have plunked down $24.95 for video coverage of the same. Others have eagerly consumed every published shred on "barn finds," project cars preserved indoors for years in a state of suspended patination before being rescued by intrepid old-car spotters.[16] At times these salvage celebrants have found themselves on the defensive, too. Sometimes this occurs when the strong arm of the law threatens one of their favorite junkyards or when it sets its sights on the project cars queued in their driveways. It also happens when corporate officials and federal, state, and local authorities join forces, as they have on a number of occasions over the last twenty-five years, in attempts to short-circuit the automotive circle of life by purging the streets of older cars and preventing their dismantling for parts.[17] In short, this third point of view on salvage yards and salvaged cars tends to see promise in patina, classics in clapped-out clunkers, and value in the rusted automotive relics others gladly cast away.

"Tends to" is a critical operative phrase here, for like the enthusiast community itself, this third narrative is anything but monolithic. Not every gearhead enjoys the junkyard experience, or actively participates in the reuse of salvaged parts through custom and restoration projects, or even cares that much about old cars. Likewise, not everyone who does enjoy "the junkyard crawl" necessarily conceives of the experience in precisely the same way.[18] Nevertheless, among hot rodders, foreign and sports-car enthusiasts, customizers, street rodders, and restoration hobbyists alike, a celebratory take on the automotive afterlife is remarkably widespread, and one or another element of it has long thrived in every corner of their world. The origin and evolution of this outlook and the practices associated with it are the subjects of this book.[19]

The first chapter sets the stage with an overview of the automotive salvage business in the United States from the 1900s through the 2000s. It builds on the work of Carl Zimring and Tom McCarthy, whose pioneering studies of the "abandoned-car problem" of the 1920s and 1950s, the large-scale recycling efforts undertaken by automobile manufacturers like Ford, the environmental regulation of the salvage industry beginning in the 1960s, and the business and technology of the scrap-metal trade are indispensable for understanding the historical scale and scope of the automotive afterlife.[20] The opening chapter of this book touches on many of these episodes and themes, but it focuses specifically on the history of the salvage-yard business itself—on the day-to-day operations of those who purchase wrecks, dismantle them in one way or another, and sell the resulting used parts to mechanics, dealers, shops, and do-it-yourself motorists. In other words, chapter 1 examines the reuse or *direct* recycling of automotive parts and systems, rather than their *indirect* recycling as a source of scrap steel. Though structured around a macro-level narrative involving two of the more dominant approaches to the business, streamlined and open-yard salvage, chapter 1 also delves into a number of more specific stories related to the business, including the roles of the insurance industry, the cycles of the used-parts market, and strategies for inventory management, regulatory compliance, and customer service. Along the way the chapter also touches on the widespread use of salvaged parts and systems among commercial repair shops and at-home do-it-yourselfers.

Against this backdrop, chapter 2 brings us to our chief concern by focusing on the outliers—those who use recycled parts and systems from wrecked cars in creative and unusual ways. To be clear, most of the parts that salvage yards recover ultimately end up back in use in humdrum ways. This is especially true of late-model engines, transmissions, and other drivetrain parts—the bread and butter of the trade—but it also applies to unbent body panels, good clean glass, and other serviceable odds and ends.[21] However, many of these same parts and systems are destined for second lives that are by no means ordinary, ending up in fine art, public works and commercial displays, or, more commonly, enthusiasts' garages. Chapter 2 examines each of these alternatives in turn. It begins with a brief survey of the "junk art" scene over the course of the twentieth century, focusing in particular on the work of those who either repurposed salvaged cars and components into sculptures, public displays, or commercial edifices, or whose work attempted to capture the in situ essence of "the junkyard aesthetic." It then explores in greater depth the reuse and repurposing of junkyard-sourced components among car enthusiasts. More precisely, it unpacks the histories of five specific groups of gearheads who have long worked closely and creatively with salvaged cars and parts: hot rodders and restoration hobbyists, beginning on a small scale in the 1910s and 1920s and as mass phenomena in the 1950s; customizers, beginning in the 1950s; import and sports-car fans, also beginning in the 1950s; and street rodders, beginning in the 1960s. Over the years, they've turned twisted wrecks into gorgeous customs, derelict cars into drag-strip terrors, seemingly hopeless piles of wood and rust into gleaming concourse restorations, and various odds and ends into street-worthy cruisers. Their final products may or may not count as "art," but there's undoubtedly an art to what they've done with others' junk.

Chapter 3 examines the commercial ramifications of these gearhead subgroups' growth. During the 1950s and 1960s, as their ranks and thus their need for project cars and specialized parts swelled, those who took part in these hobbies spawned a number of new businesses. Speed shops and high-performance firms appeared for hot rodders and other high-performance enthusiasts and, later, street rodders. Custom body, paint, and interior specialists began to cater to custom-car fans. Small reproduction and restoration shops opened up for the antique- and classic-car crowds, as did parts importers, metric-fluent shops, and other outlets for import and sports-car enthusiasts.[22] Many salvage yards also began to tailor their activities to the needs of car enthusiasts during these postwar decades, and how and why they did so is the primary focus of this chapter. Some began to hold on to their stocks of prewar cars as long as possible, and word of mouth within the restoration community made them favored destinations among those in need of obsolete parts and systems. Others actively added to their reserves of older cars and parts by buying stocks from other yards, while a number of others became specialists in one or another favored brand or era: 1920s Packards, Fords of all periods, and so forth. Many others focused instead on the mortal remains of imports and sports cars, for which many parts and systems could be difficult to locate even when the vehicles themselves were new. By the end of the 1960s there were salvage yards for every nook and cranny of the enthusiast universe, and their ranks would further swell during the 1970s and early 1980s before peaking in the later 1980s and 1990s. Chapter 3 details the origins and evolution of this enthusiast-specialty salvage sector, as well as the business practices that set it apart from mainstream yards. These include its strategies for advertising, customer access, and inventory control, as well as its gradual but ultimately overwhelming geographic concentration in the American

Southwest. Like Grand Rapids in its heyday for furniture, Detroit for new cars, Silicon Valley for software, and Los Angeles for automotive speed equipment, states like Arizona came to be the preferred source for "A-1 rust-free parts."

But even with the advent of these specialty-salvage enterprises, the search for a specific part, subassembly, or suitable project car often wasn't easy. Some makes and models were never common to begin with. No amount of effort on the part of salvage specialists could change the fact, say, that 1942 Lincoln Zephyr Club Coupes, 1949 Volkswagen Cabriolets by Hebmüller, or 1970 Hemi R/T Dodge Challenger convertibles were scarce when new and scarcer in their yards.[23] Even cars that once were common—Model Ts, Model As, Series AA Chevrolets—gradually grew fewer and farther between on salvage properties. As a result, as early as the mid-1950s the search for certain cars and parts had grown sufficiently difficult that it began to assume an almost mythical quality among gearheads. Some even began to describe their parts-sourcing endeavors as real-life "treasure hunting," complete with tall tales about backwoods finds and hidden caches of parts, as well as lamentations over the ones that got away. During the 1960s and 1970s, this treasure-hunting metaphor became a common trope among enthusiasts, and so too did a curious corollary: a widespread appreciation of junkyards *as* junkyards and of out-of-service wrecks in virtually any setting *as* out-of-service wrecks. That is, at precisely the same time that much of the rest of the country eagerly embraced measures to contain and constrain the necessary evil of the salvage business, many enthusiasts began to celebrate not just what parts they might find in a given yard but the yard itself, as well as its in situ contents. Posters and calendars featuring junkyard scenes soon multiplied, as did enthusiast-periodical coverage of treasure-hunting escapades and salvage-yard scenes. Chapter 4 delves into the lived experience of these developments. It begins by tracing the origin and evolution of junkyard treasure hunting, both in practice and as a rhetorical trope. It then explores three types of treasure hunting that did not normally take place in salvage yards but involved precisely the same mind-set: the search for unused "new old stock"(NOS) parts in every corner of the globe, the search for bargain finds at automotive swap meets coast to coast, and the pervasive mythology of the "barn find." The chapter ends with a detailed look at how the quest for automotive treasure gave rise to a widespread appreciation among gearheads for the sights, sounds, and smells of the junkyard experience.

The final two chapters zoom in on a thread first introduced in chapter 1—local, state, and federal regulation of the salvage business—and consider its ramifications for car enthusiasts as well as specialty and general-purpose salvage yards. Chapter 5 examines local developments. In particular, it focuses on a wave of local antijunkyard initiatives that followed on the heels of the federal Highway Beautification Act of 1965—efforts in neighborhoods, towns, cities, counties, and states to rein in salvage enterprises and the individual possession of out-of-service vehicles, all in the name of beautification. Since 1965, community leaders and homeowners across the country have expended a considerable amount of effort targeting long-established urban and suburban secondary-materials businesses by means of ever more onerous zoning and licensing requirements, restrictive bylaws, and the widespread abuse of eminent domain. Salvage-yard owners and automotive enthusiasts alike have long fought back against these efforts, and from time to time they have come out on top. But far more often than not, their businesses have closed and their projects have been towed away, all to make room for mixed-use zones of lofts and boutique shops in formerly industrial districts, or for clusters of pristine exurban homes with nothing but shiny new cars in their driveways and unspoiled vistas in every direction. Chapter 5 examines these urban, suburban, and rural land-use conflicts and their consequences for salvage businesses and gearhead groups alike. More broadly, it seeks to demonstrate that widespread NIMBYism has not only narrowed the opportunity for the reuse, repair, and direct recycling of the automotive past. Much more ominously, it has also helped perpetuate the fiction that a world without salvage yards is not only desirable but possible.

Chapter 6 considers the origin, evolution, and implications of a series of efforts broadly known as "accelerated vehicle retirement," "clunker," or "scrappage" programs during the 1990s and 2000s. These voluntary programs, championed by smokestack-industry executives as well as local, state, and federal politicians and environmental officials, sought to purge the streets of older cars so that their supporters could meet their federal- and state-level air-quality-improvement targets on the cheap. Put another way, they aimed to offset stationary industrial pollution by eliminating old, unsightly, and presumably gross-polluting vehicles from service, thereby hastening the pace of automotive obsolescence in the name of cleaner air and fatter corporate profits. Moreover, many of these programs sought to compound their gains by prohibiting parts recovery from the vehicles they retired, which short-circuited the salvage industry's normal routines and effectively accelerated the demise of other older models still in daily use. The goal, in short, was to send these superannuated vehicles straight from the street to the crusher and from the crusher to scrap-metal processors. For salvage yards and automobile repair shops, accelerated vehicle retirement programs were therefore a direct threat. For enthusiasts, especially but not exclusively antique- and classic-car enthusiasts, these programs were also menacing both because they tended to bypass salvage yards and because they indiscriminately targeted the older cars that many gearheads cherished. For a number of environmental and social-justice advocates, scrappage schemes were problematic for a host of other reasons, not the least of which was the assumption that getting rid of a certain number of so-called clunkers made it acceptable for corporations to continue to pollute. Vigorous opposition from these varied interest groups ultimately led to the abandonment of accelerated vehicle retirement programs in all but a handful of areas. But as this final chapter demonstrates, over the course of its twenty-odd-year run as a widely embraced policy, "scrappage fever" reignited fundamental debates, once thought long-settled, over the relationship between the old and the new in the modern American economy. A brief conclusion follows, which explores the role of salvaged cars and parts in ever-shifting gearhead attitudes toward novelty, obsolescence, and the automotive past.

Through it all, *Junkyards, Gearheads, and Rust* aims to make three broad contributions. First and foremost, it seeks to further flesh out our understanding of what happens to consumer durables like cars at the end of their useful lives. More specifically, it aims to fill a gap in the literature on the automotive afterlife by emphasizing the activities of those who interact with cast-off cars after they are retired from use but before they are sent to the shredder and the scrap-metal furnace: ordinary and specialty salvage yards, used-parts dealers, and car enthusiasts. In so doing, this book sheds light on the many ways in which creative reuse and direct recycling have thrived throughout the automotive age, an era ostensibly infused with an obsolescence mind-set and a throwaway mentality. For if nothing else, gearhead activities like customization, street rodding, and restoration clearly suggest that bricolage, which Douglas Harper, Susan Strasser, and others have lamented as a dying art, is and has long been alive and well within the automotive realm.[24]

Second, by exploring automotive bricolage among junkyard-scouring enthusiasts, this book also speaks to David Edgerton's critique of the history of technology as a field that pays too much attention to exceptional moments of novelty rather than the more mundane but no less vital routines associated with technology in use.[25] Consider our car culture. Much to the chagrin of the automotive industry, American streets have long been dominated not by shiny, brand-new cars, but by models many years and miles their senior, sold and resold several times and kept in service via parts depots, repair shops, and other means, including salvage yards. Building on Kevin Borg's work on the repair business and Steven Gelber's on the used-car trade, *Junkyards, Gearheads, and Rust* seeks to broaden our understanding of this world of car technology in use.[26] Its focus on enthusiasts and salvage yards may seem somewhat off the beaten path, at least at a glance. But custom cars, hot rods, street rods, sports cars, and restored antiques have long been central elements of the American car culture. And as this book shows, their origin and evolution as distinctive hobbies hinged in large part on the growth and maturation of the automotive salvage business.

Finally, and most broadly, this book weighs in on an intermittent dialogue about obsolescence and the American consumer that has been developing for many years. Among social critics and public intellectuals, this includes Vance Packard's postwar trilogy on producer and consumer profligacy; John Keats's charge that American consumers were being duped into buying needlessly lavish "insolent chariots"; critiques of the wasteful postwar urban and commercial landscape by both John Kouwenhoven and Peter Blake; and Kevin Lynch's ruminations on the nature and meaning of waste itself and the inherent follies of both wasteful and so-called waste-free living.[27] It also includes efforts to "close the loop" by advocating an economy centered on recycling and materials recovery, as well as a number of more recent exposés on the end results of the obsolescence-oriented economy in which we actually live: mountains of garbage and cast-off consumer durables.[28] Among anthropologists, historians, and others, this intermittent dialogue includes serious attempts to grapple with the actual

as well as the imaginary problems associated with our throwaway economy; useful analyses of the social and economic dynamics of the trade in
used and collectible goods; and efforts to understand the scale and scope
of the garbage, recycling, and secondary-materials industries.[29]

Missing in much, but by no means all of this scholarship are serious
considerations of those for whom the old and obsolete are no less desirable
than the new and up to date. Among the exceptions are a handful of essays
in the field of media studies that examine the afterlife of nominally obsolete
technologies—vinyl records and out-of-date computer systems—among
dedicated high-end audiophiles, vintage gamers, and electronics hobbyists. These enthusiasts do not necessarily reject all new media technologies out of hand, but they also do not wish to reflexively abandon their old
ones simply because newer iterations have debuted. For they have found,
in a manner reminiscent of the work of Leo Marx, that technological change
does not necessarily result in "progress." They have realized, that is, that
in terms of the qualitative experiences they seek, newer media forms
and devices do not necessarily deliver valuable improvements.[30] Their
nuanced—dare I say, *healthy*—approach to obsolescence and consumption
is of course quite rare, especially in the realm of media and information
technology.[31] But as this book demonstrates, a strikingly similarly approach
to the old and the new has also long been common among car enthusiasts.
To be sure, much of the extant literature on obsolescence and waste tends to
point in precisely the opposite direction, toward the conclusion that the
American car culture is the very last place where one might expect to find
a nuanced approach to the throwaway mentality.[32] After all, it is against
the wasteful postwar automobile industry that Packard directs a good deal
of his outrage, and Keats the entirety of his. It is the automobile industry
that gave us the annual model change, too, selling the American public on
the promise of an ever-better four-wheeled future and ultimately leaving
it with untold millions of unwanted cars in roadside junkyards. Long the
poster child of the throwaway economy, the automobile certainly seems to
be the quintessential good that is by definition destined, in the words of
the social anthropologist and rubbish theorist Michael Thompson, for the
oblivion of the scrap heap.[33]

This book seeks to complicate this narrative about obsolescence and
the automobile. It does so not by working to deny the living legacy of Al-
fred Sloan's approach, for the bulk of today's car-buying public is no less

sold on the allure of the new than it was when Calvin Coolidge lived on Pennsylvania Avenue. Instead, *Junkyards, Gearheads, and Rust* aims to complicate our understanding of the role of the old and the new in American automotive history by delving into the attitudes and practices of those who give that history so much of its color and flair. For the world of the hot rodder, customizer, street rodder, import and sports-car fan, and restoration hobbyist is not one in which each new model year necessarily represents the proverbial "last word" in engineering, performance, and style. Instead it is a place where brand-new cars are routinely hacked apart and rebuilt to better suit their owners' tastes. It is a place where many will refuse to own a new car but will gladly drive an old one retrofitted with state-of-the-art equipment. It is a place where the most modern of chemical and metallurgical techniques are routinely applied to the resurrection of long-obsolete antiques. It is a place where the prospect of being stranded on the side of the road by the untimely failure of a superannuated Lucas or Bosch component is but part of the fun to some and sheer lunacy to others. It is a place where well-stocked salvage yards aren't unwelcome eyesores but vital wellsprings of creative energy. It is a place where those who love cars often rail against the follies of the automotive industry, its products, and its customers, but just as often rally to their defense.[34] It is also a place in which the automotive industry itself has long found heritage-marketing opportunities and grist for their retro-styling mills. For when it comes to the old and the new, the world of the enthusiast is refreshingly gray, a liminal space in which the past and present come together in unusual but instructive ways. At the epicenter of this creative space is the automotive salvage yard, the gearhead's favorite playground and the primary focus of the pages which follow.

So put on your boots and your coveralls, and grab your toolbox. It's time to wander in among the weeds and rust.

The Automotive Salvage Business in America, 1900–2010

An Overview

In the spring of 1999, the Ford Motor Company announced that it was entering the North American automotive salvage business. This was not the first time that the company attempted to bring its economies of scale to the sector. The firm once operated a large-scale disassembly and salvage program at its Rouge facility, a money-losing venture launched in the 1930s that the firm abandoned in the 1940s. Nevertheless, top Ford brass were optimistic that this new endeavor would go more smoothly and be longer lived. Part of CEO Jac Nasser's "grand plan to get more of the dollars from a vehicle's entire life," the strategy was simple enough: enhance the firm's spare-parts business by assembling a network of wholly owned salvage yards across the continent, each of which would operate in a streamlined and environmentally responsible fashion. At each yard, incoming vehicles would be dismantled rapidly, salvageable parts and systems from them would be inventoried and cleaned or rebuilt as necessary, and the remaining stripped-clean shells would be sold for scrap. Dealerships, repair shops, and individual customers would be able to search the network's inventory via the Internet and rapidly access the parts they needed without ever setting foot in a salvage yard. Operated by a new subsidiary of the Ford Motor Company known as GreenLeaf Acquisitions, Nasser's effort ultimately brought together more than thirty salvage yards and quickly won the firm a lot of positive coverage in the automotive, information-technology, waste-management, and mainstream press. Unfortunately for Ford's investors—if not for Nasser, whose short-lived tenure as the head of the company ended in 2001 for other reasons—GreenLeaf also failed to turn a profit. More precisely, if less charitably, it turned more than thirty independent and profitable businesses into a network of thirty dependent and money-losing businesses. GreenLeaf also brought the Dearborn giant major legal headaches by failing to adequately distinguish its salvaged and rebuilt components from its brand-new replacement-parts offerings. By mid-2002, Ford's top executives were prepared to wash their hands of the affair, and in 2003 they sold off Green-Leaf Acquisitions.[1]

Apart from Ford's direct involvement, GreenLeaf's approach to automotive salvage embodied nearly every cutting-edge trend in the business at the time. Several other large-scale companies—Atlanta-based Pull-A-Part, for one, as well as Chicago's LKQ Corp. and San Mateo's Copart—actually thrived during the 1990s and 2000s by doing precisely what GreenLeaf attempted: assembling national networks of wholly owned yards.[2] Also, many salvage businesses of every size had long since shifted to computerized record keeping, and a good number were beginning to experiment with Internet-accessible inventories as well.[3] More broadly, the very fact that Ford decided to try its hand at automotive salvage reflected a growing interest among 1990s automakers and policy makers in creating a more recyclable car. Though much of this recyclability initiative was centered in Europe, Ford's Dearborn-based engineers were actively engaged in materials-recovery research throughout the decade, and the company's decision to move into the salvage business in 1999 must be seen against this broader backdrop.[4] Finally, and most fundamentally, the way in which GreenLeaf's networked yards approached the task of actually salvaging parts from their wrecks was entirely in sync with the most salient trend of the time, a shift toward streamlined operations. Paul D'Adamo, the owner of a Cumberland, Rhode Island, salvage yard, explained the trend quite well to the *Providence Journal* in 2007: "We are not letting cars sit around in the yard as much as we used to. . . . The idea now is to get the cars in and get them out fast. We're like a factory, but in reverse."[5] D'Adamo's business still relied on temporary outdoor storage, but other salvage businesses in the 1990s and 2000s dispensed with the yard altogether. Instead, these firms dismantled their wrecks, warehoused the resulting parts, and sold the remains for scrap more or less as soon as they arrived.[6] Not every yard attempted to streamline its operations in these ways, and plenty of old-school yards with low turnover and an inventory-by-memory approach still thrived. But the headline-grabbing trend at the end of the twentieth century and the beginning of the twenty-first was abundantly clear: system and order in place of weeds, mud, and rust.

The trend was real, and to those within the business and those reporting on it, it certainly seemed to be a genuinely novel approach to automotive salvage. But it wasn't. Back in the late 1900s and the early 1910s, when dedicated wrecking yards first began to appear, nearly all of them approached the business in a streamlined manner not unlike the headline-grabbing methods of a century later, complete with rapid turnover, inventoried and warehoused parts, and squeaky-clean retail counters. Only when the tide of unwanted cars began to overwhelm the business in the 1920s and 1930s did open yards with heaps of rusting wrecks become the norm. So it would remain until the late 1980s and 1990s, when a number of yards coast to coast began to adopt the more streamlined approach described by D'Adamo in 2007. In the narrowest sense, few if any individual 1990s and 2000s yards actually *reverted* to such an approach; in their owners' eyes, and in the eyes of the press, they were developing an altogether *new* strategy for the dismantling of wrecked or otherwise unwanted cars. But in a broader sense, the salvage industry as a whole was simply circling back.

"We Tear 'Em Up and Sell the Pieces"

When it happened at all, automotive salvage was a nebulous activity at the dawn of the auto age. Established junkmen and scrap-metal dealers and processors handled some of the castoffs of the 1890s and early 1900s. In their hands, some were broken up for parts, while others were processed as scrap and sold wholesale to large-scale secondary-materials brokers acting on behalf of the steel industry.[7] Others avoided the scrap-metal stream altogether, ending up in pieces at flea markets, where creative tinkerers would hunt for odds and ends,[8] or behind blacksmith, bicycle, and other repair shops. There they gradually yielded parts to mend other cars or were cannibalized for their engines, which went on to serve as stationary power plants in shops or on farms.[9] Other castoffs remained intact and often in use. Some were sold off for continued service in smaller towns, rural areas, or markets outside the United States, while others were tucked away in stables and barns when their wealthy, early-adopter owners moved on to newer models. There they would remain for months, years, and even decades in a state of extended limbo, neither sufficiently useful to remain in service nor useless enough to be scrapped.[10]

Their fates were varied; less so the reasons they were discarded in the first place. Some were simply damaged beyond repair in collisions. Accidents involving self-propelled vehicles were nothing new, even at the dawn of the automobile age. In 1771, steam-carriage pioneer Nicolas Cugnot is said to have driven his *fardier à vapeur* into a wall in Paris, and in the 1860s Richard Dudgeon reportedly ran his experimental steam wagon into a barber's shop on Long Island. Neither event is particularly well documented, and Cugnot's misfortune in particular may well be the stuff of legend.[11] Quite real, on the other hand, were municipal efforts to constrain horse-drawn vehicles, steam-powered omnibuses, and ultimately horseless carriages of all sorts because of their propensity to create mayhem on nineteenth- and early-twentieth-century streets.[12] Although horseless-carriage boosters at the time derided these regulatory efforts, the statistics in the earliest days of the automobile age were indeed quite grim. As early as 1899, when there were only eight thousand road-going motor vehicles of any sort in the United States, twenty-six people lost their lives in traffic accidents, and by the time the Model T was introduced, the annual figure was well on its way to quadruple digits and would soar past ten thousand by the end of World War I.[13] Some of these fatalities involved collisions with pedestrians, while others were the result of single- or multiple-vehicle crashes. Either way, many of the cars involved were "totaled"—that is, they were declared unrepairable (or at least economically unrepairable) "total losses"—either by the owners themselves or, in increasing numbers as automobile ownership and use spread, by the insurance companies with which their owners held protective policies.[14]

At least as common was the practice of discarding automobiles either when they ceased to function as expected or when they appeared for other reasons to be obsolete. The cars of the 1890s and the early 1900s required frequent maintenance, and many, especially those at the lower end of the market, had very short life expectancies before major repairs became necessary. "Most of us whose motoring experience extends back more than five years," *Motor*'s Herbert L. Towle remarked in 1910, "can recall many cars whose owners were glad to sell them for scrap-iron at the end of a single, brief and checkered season's experience." Automotive reliability improved a great deal during the mid- to late 1900s, but even these later vehicles would have been lucky to escape their second or third years of service without major complications. Assuming their owners performed the maintenance and repairs required, most of these cars had the potential to sputter on and on and on, providing them with many years of service. But instead, as Towle explained, lower-priced cars tended to "change hands at the end of the first season, the first owner skimming the cream of their usefulness, and passing on to his successor the job of overhauling and the more or less frequent tinkering required by a car that has begun to wear out."[15] By getting rid of them so soon, first owners not only avoided the financial burden of major repairs, but they also escaped the economic trap of throwing good money after bad. Automotive technology developed rapidly during the decade of the 1900s, rendering any car more than one or two years old functionally obsolete—"ancient," as the motoring press often put it.[16] By disposing of them rather than paying for major repairs to keep them on the road, these first owners were able instead to spend their money on newer and, in most cases, vastly superior cars. Doing so also allowed status-conscious first owners to avoid the embarrassment of being seen in out-of-date models. On the other hand, the subsequent owners of these discarded cars were typically lower on the socioeconomic scale and often lived in small towns, rural districts, or foreign markets where there was "no great premium on the latest models." At a time when maintaining a used car was still cheaper than buying a new one, second and third owners less concerned about appearances often squeezed many years of service out of their ostensibly obsolete vehicles. Such was the fate of most cars first sold in the 1890s and early 1900s.[17]

Total registrations rose quite quickly during this early period, as did traffic fatalities. But not until the later 1900s did the resulting wrecks and other automotive discards that could not be resold reach a sufficient critical mass to support salvage businesses specializing in automobiles. Even then, the going was tough for those who tried to make a living by "tear[ing] 'em up and sell[ing] the pieces."[18] The Auto Salvage and Parts House of Chicago—according to the best available evidence the first firm organized specifically to handle automotive salvage—opened its doors in 1907 with an inventory of twenty-eight wrecks, a small yard, eight hundred square feet of indoor space, and two employees. The plan was to dismantle these cars (or "wreck" them, as it was then described), clean the resulting parts, store those that were still serviceable in inventoried bins, and sell them to repair shops for reuse on other cars of the same make. In its first year, sales were pitifully slow, possibly because all twenty-eight of its first wrecks were high-end, low-volume Pope-Toledo cars, and according to the firm's president, "it was sometimes hard to discover sufficient profit to pay for three meals a day." But as the age of mass automobile use dawned, the company's fortunes improved dramatically. During the early 1910s its burgeoning incoming inventory necessitated several successive expansions of its outdoor wrecking yard and indoor storage spaces. By its tenth anniversary, the Auto Salvage and Parts House occupied 12,000 square feet of interior space, dismantled 150 cars a year, and had whittled its process down to an average of 15 hours per car, including cleaning, sorting, and labeling the resulting parts.[19]

The key to the success of the Auto Salvage and Parts House was a robust turnover of incoming cars and outgoing parts. In other words, this was not a stagnant automobile graveyard where cars were stored for years on end, gradually yielding parts. Instead it was a streamlined operation that plucked the most marketable parts from its incoming cars and sold the rest as scrap. Those in need of parts did not have to fumble among the wrecks, and they did not even need to visit the operation's sales counter: early on, the Auto Salvage and Parts House incorporated mail orders into its streamlined business model.[20]

The same model was pursued by most of the other wrecking firms, numbering approximately thirty, that entered the business prior to World War I. Consider the Guy Auto Exchange of Montclair, New Jersey. With ten dollars in capital and six hundred square feet of ramshackle indoor space, this firm opened its doors in 1907, just after the Auto Salvage and Parts House of Chicago. It too sought to profit by dismantling incoming cars, warehousing the resulting parts for retail sale, and scrapping the remains. It too grew dramatically in the 1910s as well, from an average of one to two cars processed each month to more than thirty by early 1917.[21] The same was true of the Auto Salvage Company of Kansas City, Missouri. Its operations commenced in 1911 with three employees and a single floor of urban space. By focusing on the streamlined dismantling of middle-class models, it was able to both expand its primary location and establish satellite operations in Cincinnati and St. Louis. By 1917 it was wrecking thirty to thirty-five cars each month across its three large shops. This allowed it to maintain an impressive inventory of parts, "having on hand, for example,

Figure 1. The dismantling area of an early salvage yard, the Auto Wrecking Company of Kansas City, Missouri, pictured in 1919 upon the occasion of its first foray into aircraft salvage. (*Motor Age*, 9 April 1919, p. 21.)

connecting rods for sixty-two different makes of cars." Also, its dismantling and inventory process was sufficiently well organized to publish a monthly catalog, through which repair shops and interested owners coast to coast could order the parts they needed through the mail at an average price 50 to 75 percent less than new.[22] Perhaps the largest and best-known of these early streamlined firms, however, was the Auto Wrecking Company of Kansas City, Missouri. From humble origins in 1913—its initial inventory consisted of a single fire-damaged Buick Model F—it grew rapidly, wrecking 150 cars per month by 1917 and, beginning in 1919, the occasional airplane as well (fig. 1). Although it was the first to exclaim in its advertisements that "We tear 'em up and sell the parts," the Auto Wrecking Company did not deal exclusively in parts from dismantled cars. Through an East Coast buyer, the firm also purchased inventories from bankrupt dealers and other new-parts depots, supplementing its well-sorted inventory of used parts and enabling it to fill 85 percent of the mail and counter orders it received from stock.[23]

Not every firm established in this period was quite as well organized, at least at first. The Auto Salvage and Exchange Company of Des Moines, for example, opened its doors in late 1915 and proceeded to dismantle its incoming inventory without paying much attention to how the resulting parts were sorted and stored. Within six months, its modest shop was crammed to the brim with piles of parts, and its managers quickly realized that the resulting "haphazard searching for gears or axles in the great heaps was causing a loss of time and money." By 1917, according to a contemporary description, the firm had thoroughly reorganized its operations and become a model of streamlined efficiency in its inventory and storage processes:

> It has had several blank forms printed, which display all work performed and stock on hand. One blank is given to the wrecking foreman at the time of the delivery of a car to him. It provides a complete history of the condition of the car when disassembled, lists all the various parts taken from it and contains a separate column where all parts found useless are classified as junk. Another blank is used for stock listing, and since it records every article placed in stock and is indexed and filed, the company is able to discover quickly just what parts it has on hand and is in a position to take care of rush and long distance orders promptly.

The firm also maintained what it called an "emergency car" for same-day deliveries on local rush orders placed by phone, although the bulk of its business, like most of its peers', was handled over the counter or by mail.[24] Others added their own twists, especially in their advertising. Some emphasized prompt shipping, while others stressed the range of makes and models for which they had used parts in stock and ready to ship. Some offered money-back guarantees, and nearly all were keen to underscore the savings one could realize by buying used instead of new. At least one was eager to distance itself from any association with "junk" shops—"Don't get the idea that this stuff is 'junk,'" explained one of its advertisements from 1916, "'SALVAGE' means serviceable and 'SALVAGE' is our middle name"—while the owner of another actually embraced the association, proudly telling *Motor Age* in 1917 that "I'm the guy that put junk in the junk shop."[25] Far more important than any of these minor differences in marketing and sales, however, were the basic traits these salvage companies shared. They were urban, giving them ready access to an abundant supply of wrecked cars and used-parts customers. They were also interested less in the scrap value of the iron and steel in a given wreck than they were in the cumulative value of its salvageable parts. Finally, and most critically, their operations were streamlined in approach, with high rates of turnover among their incoming queues of wrecks and rapidly rotating stocks of parts on hand.

A similar approach was also used extensively during World War I by the major powers on both sides. Early in the war, one hundred miles to the west of the front in France, the British military established a large salvage and repair depot. Vehicles of all sorts damaged in the conflict—chiefly British-, French-, and German-made trucks, small runabouts, and large touring cars—were gathered and sent by rail to this facility. There they were sorted in a large receiving yard into neatly aligned rows, with each car marked as destined for either repair or dismantling. "Some are marked down as scrap," a contemporary account explained, "but it is very rarely that an entire unit has to be sacrificed." Instead, those damaged beyond repair were normally dismantled and their serviceable parts warehoused indoors, not unlike a typical American salvage business of the time.[26] Even more streamlined was France's central repair depot, where those in charge "developed automobile dissection into a fine art" in which every piece of an irreparably damaged vehicle was scavenged and reused. Engines, radiators, tires, and other major components were removed, cleaned, and warehoused; interior fabrics were sorted for processing and reuse; and any metal that could not be directly recycled back into use was sorted by type, gathered into heaps, and sent off to be smelted down.[27] Not to be outdone, the Germans established a similar depot at Tempelhof in Berlin, and the American military's Motor Transport Corps (MTC) set up a massive facility on 337 acres 120 miles behind the front in France in early 1918, variously known as the "Salvage Park" or the "Motor Reconstruction Park." Both salvage and repair work took place at this MTC depot, which employed 3,500 skilled workers from the American automobile industry and 1,800 German prisoners and civilians. Like the British, French, and German facilities, the MTC depot cleaned and warehoused parts from vehicles beyond repair and used them to rebuild others in less dire straits. Those familiar with the American salvage firms of the era would have felt right at home at any of these facilities, save for their scales of operation and the wartime nature of the automotive damage found within them.[28]

Back home, things began to shift a bit after the war. An increasing number of salvage companies rounded out their retail offerings by adding new replacement parts to their inventories. Others emphasized their ability to duplicate hard-to-find items in on-site machine shops.[29] Tow trucks became common as well. First developed by a Chattanooga, Tennessee, mechanic by the name of Ernest Holmes in 1916, who patented the idea in 1917 and went on to manufacture them for many years, tow trucks made the job of gathering incoming salvage inventory substantially easier for those firms that adopted them. Some yards built their own by modifying trucks and other vehicles, while others bought them from specialty manufacturers like Holmes.[30]

But by far the greatest shift within the business after World War I came about because of shifting market circumstances. Most salvage firms still aimed to rapidly dismantle their incoming wrecks and warehouse the resulting parts in the early 1920s,[31] but adhering to this streamlined model was becoming increasingly difficult. For as the new-car market reached its saturation point, the number of older models scrapped each year rose dramatically, from just under 275,000 in 1920 to more than one million by 1924. So did the number of cars that were simply abandoned at the curb or in vacant lots.[32] As the volume of available wrecks mounted, the number of urban salvage firms grew, and by the middle of the 1920s they were a common sight in cities coast to coast. But even as the sector expanded, the individual firms within it fell further and further behind because the fundamental challenge of the day was one of productivity, not scope. Simply put, the established process of wrecking cars by dismantling them, sorting and storing their useful parts, and burning off the remaining wood and fabric, all by hand, before hauling the rest to the scrap yard was far too cumbersome to keep up with the rising tide of unwanted cars.[33] Consequently, many firms' actual wrecking or salvage yards—the outdoor spaces where damaged cars were briefly stored upon arrival before being dismantled and where the dismantling process itself often took place—began to grow, as did the volume of unprocessed cars and parts within them. By default if not design, the "open" style of yard was gaining ground.

Pioneered in 1915 by Morris Roseman of Lancaster, Pennsylvania, the open style of wrecking operation still involved organized dismantling, and it still aimed to maintain a rapid pace of turnover. But in a business of this sort, a large proportion of incoming wrecks, principally those with lots of serviceable parts, were stored intact in open, outdoor spaces rather than being dismantled straightaway into thousands of tagged and warehoused parts. This enabled a greater variety of components to be salvaged from each wreck, including oddball parts that streamlined yards routinely discarded, and it helped cut labor costs, too. It also introduced to the salvage business the practice of allowing customers to browse through inventories of yet-to-be-dismantled wrecks. However, it also meant the emergence of salvage operations that deliberately sprawled out over several acres, their wrecks "generously distributed over every available space," as one contemporary described an open yard in Springfield, Massachusetts, in 1921, "leaving here and there a few alleyways over which the prospective customer or curiosity seeker may find his way."[34] This, coupled with outdoor inventory backlogs at nominally streamlined operations, which were less and less able to keep up with rising annual scrappage rates, meant that outdoor storage soon became a common feature of the industry rather than an aberration. As a result, by the end of the 1920s the horizontal and vertical expansion of the typical yard—literally speaking, as salvage acreage itself and the stacks of cars upon it grew—began to draw fire from those concerned with urban blight.[35]

In spite of the emergence of sprawling and unsightly yards, and in spite of the burgeoning number of cars that were scrapped each year—more than two million by 1928 and close to three million by 1931[36]—streamlined dismantling was not yet a thing of the past. In fact, two trends during the 1920s and 1930s actually gave it a renewed lease on life. More precisely, if somewhat paradoxically, these trends helped to sustain the streamlined approach *precisely because* of the rising tide of unwanted cars. The first involved the nation's automotive retailers. As early as 1911, dealers in larger cities began to complain about the number (and prices) of used cars coming back in partial trade on new models. They did not want to stop taking these trade-ins altogether, for doing so might well scare off potential new-car customers who already owned earlier models. But dealers also did not want to continue to take those older cars on trade if they could not reliably resell them at a profit.[37] By the middle of the 1910s, this "used-car problem" had become sufficiently acute to prompt the deployment of experimental solutions. Some rebuilt unwanted used cars into sporty speedsters for easier resale, swapping their fenders and hammering out new body panels aft of the cowl. As *Motor Age* explained in 1921, however, the resulting cars were often "not quite distinctive enough to appeal to the best trade," leading some to favor ready-made speedster bodies bought from aftermarket firms. Others instead converted some of their used cars into light-duty trucks for commercial and industrial applications, a straightforward operation similar to the construction of a speedster. A few experimented with consignment-only used-car policies, offering to sell their new-car customers' older models for them without assuming the up-front risk and cost of accepting them in trade. Others joined forces to host citywide used-car shows to clear their lots.[38] But the problem only deepened, and during the early 1920s it would continue to worsen as the mass market for new cars reached its saturation point.[39] In response, some dealers redoubled their efforts to clean up and refurbish their trade-ins to improve the odds of profitable sales; others tried gimmicks such as used-car raffles, and dealerships in several cities also set up cooperative used-car lots to better spread the risk of accepting unwanted cars in trade.[40]

But an Overland, Hudson, and Cadillac dealership in Pella, Iowa, went further than most. Rather than trying to resell each and every used car that it took in trade, it developed a system in the mid-1910s whereby incoming used cars were separated into two streams according to their overall condition. One led to a department that refurbished serviceable cars for resale. The other led to a separate department that systematically dismantled the more worn-out trade-ins and cleaned, labeled, and inventoried their parts for resale before sending the remains to the scrap yard. In effect, the Pella Motor Company ran a streamlined, in-house wrecking operation as its own solution to the used-car problem. *Motor Age* gushingly praised the Pella dealer's system in a lengthy write-up in 1919,[41] and several years later *Automotive Industries* advocated a similar idea that was first put forth by the National Team and Motor Truck Owners' Association in 1923. In order to make the streets safer—and, more to the point, to protect the reputations and profitability of "legitimate" trucking companies threatened by fly-by-night operators willing to use cheap and worn-out trucks— the association called upon its members to consider scrapping their trucks at the end of their useful lives rather than seeking to trade them in. *Automotive Industries* and *Motor Age* jointly called for the destruction of wornout models by new-truck dealers in 1923, and in 1924, the editors of *Motor Age* ran another feature on a new-car dealership that systematically dismantled its used cars and published an editorial that openly advocated the destruction of worn-out trade-ins of all sorts by all new-car dealerships. Its logic?

> Every day we see automobiles operating on the streets that should be junked. They are old and worn out and a positive danger to the occupants of the car and to other traffic. They wobble and sputter and wheeze. . . . It is not right for the dealer to sell a second-hand car that obviously is not fit for road service. Apparently he has made a good sale, but actually he has put a defective mechanism into the hands of an owner who very likely is neither morally nor financially responsible. And having done so he has smirched his own business and the whole industry.

Instead, for the sake of public safety, the reputation of the nation's automotive retailers and, not least, the profitability of those retailers in the face of the used-car problem, unfit cars should all be junked.[42] Not until the used-car problem reached a deeper level of crisis in the mid- to late 1920s, however, did broader swaths of the trade embrace the idea.

The first to do so were the new-car dealers of Omaha, Nebraska. In early 1927, through the Omaha Automobile Trade Association, they incorporated a jointly owned wrecking yard known as the Cooperative Salvage Company. When it opened its doors that March, area dealers began to send their less-desirable used cars to this operation for dismantling. Their parts were sorted for either resale or scrap, and those deemed worthy of resale were chemically cleaned, inventoried, and stored in bins. Although its dismantling operation and some of its storage bins were located outside in a fenced-in yard, this was not an open automobile graveyard. Instead, like the pioneering salvage yards of the 1900s and 1910s, it was a streamlined dismantling and used-parts operation.[43]

Dealers in a number of cities followed Omaha's lead over the next several years. Twenty-two members of the Kansas City Motor Car Dealers Association, for example, raised $50,000 in working capital and established a jointly owned firm called the United Auto Wrecking Company in the summer of 1927. Their motive was not simply to avoid direct losses on trade-ins. Instead, these dealers also aimed to cut the Kansas City area's other salvage businesses out of the used-car trade. At the time, more than one-third of the worn-out vehicles that the city's automobile dealers sold to area salvage yards for dismantling "eventually came back in trade," meaning that Kansas City's salvage yards refurbished quite a few of the cars they processed. The United Auto Wrecking Company aimed in part to curtail this practice by ensuring that unwanted used cars actually were dismantled.[44] The same was true of the Auto Trade Salvage Corporation, established by the Milwaukee Automotive Dealers Association in 1927, as well as the Auto Dealers' Salvage Company of St. Louis, set up in 1928.[45] Dealer associations in Detroit, Sioux City, San Francisco, Cleveland, and several other cities launched cooperative arrangements of their own in the 1920s. Most followed Omaha's lead and ran their own dismantling operations, while others teamed up with established scrap processors in order to minimize their capital requirements.[46] The majority of those that operated their own dismantling facilities also sold used parts from their wrecks, but a few of these cooperatives scrapped their cars entirely in order to help boost their replacement-parts sales.[47] Some of these cooperative arrangements were profitable, at least at first, but most of them died out during the Great Depression as the market price for scrap iron plummeted, new-car sales tanked, and fewer worn-out cars were taken off the road.[48] What matters for our purposes, however, is that these dealer-cooperative salvage yards were run not as open graveyards, but as fully streamlined operations, complete in several cases with standardized price lists, used-parts catalogs, and a healthy volume of mail-order business.[49]

The second trend of the 1920s and 1930s that helped prolong the life of the streamlined-salvage model was the entry into the wrecking business of two new-car manufacturers. The first was Chevrolet. Championed by company president William S. Knudsen, its plan, launched in 1927, granted allowances to franchised dealers so that they could scrap a certain proportion of the used cars they took in on trade. Specifically, the automaker agreed to pay its dealers thirty-five to fifty dollars to junk one used car for every ten new ones sold. Each doomed car was scrapped entirely, without the recovery of used parts, and to this end the actual dismantling was done "under supervision of a member of the Chevrolet field force." The plan was popular among dealers, and *Motor Age* and *Automotive Industries* praised the automaker for doing its part to solve the used-car problem.[50]

But in the grand scheme of things, Knudsen's scrappage program was small potatoes compared with what was going on at Ford. In February 1930, a massive disassembly operation began at its Rouge facility, where trade-ins of all makes from Ford's Detroit-area dealers were systematically dismantled. According to a contemporary description, "On the salvage line

everything in [the cars] is reclaimed to serve some useful purpose. Artificial leather is made into aprons, upholstery goes into hand pads, floor boards
serve as crate tops, glass is used for window panes, and metal is utilized in
the making of steel."[51] By the end of 1930 the operation was scrapping an
average of six hundred cars per day, but its success depended entirely on
the price of its chief product, scrap steel. And when prices for this com-
modity plunged in the early years of the Depression, the disassembly line
was slowed considerably. According to Tom McCarthy, Ford then ran the
operation intermittently until 1943 or 1944, "at which point the company
abandoned it entirely."[52]

Ford's mixed experience with its disassembly operation is not altogether surprising. For with the exception of the wrecking yards run by
dealerships and the limited involvement of Chevrolet, the firm's attempt
to bring assembly-line efficiency to automotive salvage was decidedly out
of sync with the times. Instead, the overarching trend within the industry
during the 1920s and 1930s involved a shift away from streamlined operations, with their elevated rates of turnover and their indoor storage areas,
and toward expansive outdoor yards in which wrecks sat intact for months
and even years on end. Open yards of this sort had been on the rise since
the early to mid-1920s, when the volume of incoming wrecks first began to
overwhelm the capacity of some dismantlers. But especially in the context
of a depressed economy in which the price of scrap materials was low and
fewer cars were junked each year, the incentive to rapidly recover useful
parts from each incoming wreck in order to make room for the next—and
to make a quick buck off the remnants at the scrap yard—further withered. Outdoor inventories grew, and by the time the United States entered
World War II, the wrecking and dismantling business had largely given
way to open and expansive salvage yards.

"Shop JUNKYARDS and Save"

During the war, the War Production Board's Bureau of Industrial
Conservation and the Office of Price Administration worked closely with
officials in the scrap-metal trade to identify and purchase, by force if necessary, dormant "hordes" of ferrous materials, including stocks of old
wrecks in automobile graveyards.[53] But the return of peace and the recon-
version of the wartime economy gradually led to a resumption of the prewar pattern. Owing in part to the Korean War but also to the needs of a

rapidly expanding economy, the price of scrap steel remained high through the early 1950s, and this briefly kept salvage-yard inventories in check.[54] But the price of scrap soon fell, at precisely the same time that middle-class Americans began to trade in their old cars for new ones at a rapidly accelerating pace. "Old" was a relative term, of course. By 1955, the average age of the cars on American streets stood at just under six years, having fallen from an immediate postwar high of nine that was largely due to the wartime suspension of production. There it would remain, at just under six, through the rest of the 1950s and 1960s. This, coupled with rapidly rising total-registration figures (seventy million by 1966, up from thirty million just before the war), meant that a lot more cars—record numbers, in fact—were being scrapped each year after the war than before: four million by 1956, eight million by 1970.[55] Low scrap prices during the late 1950s and 1960s meant that many of these cars sat in open salvage yards far longer than they would have during or immediately following the Second World War or during the 1920s.[56]

Soon the country's tens of thousands of roadside junkyards were supersaturated with wrecks, sometimes in unsightly stacks five or six cars high, sometimes side-by-side on sprawling acres. Whenever possible, many of their operators still pulled valuable items like running, low-mileage engines from their wrecks for indoor storage. Many also set up outdoor racks to store front- and rear-end body clips, door shells, hoods, rear axles, and other components pulled from cars already sold for scrap. Thus, like those of the 1910s and 1920s, these postwar yards still aimed to profit by picking each wreck clean and selling the rest as scrap. The difference was that most no longer sought to fully tear apart each car as it arrived. Instead they stashed their wrecks outdoors intact, pulling and selling parts from them on an on-demand basis. Some did so in a manner comparable with prewar practices, employing teams of men to fill each order as it arrived. But many others adopted the simpler, less costly "you-pull-it" approach. Customers of this type of salvage yard would arrive in person, tools in hand, and check in at the front office to explain their needs. Assuming they had vehicles in stock with the parts required, the operators would then usher or point their customers to the right section of the yard and leave them to pull the parts themselves. Over time this business model came to be the norm, and throughout the 1950s, customers who were "at all handy with a screwdriver and kn[e]w a wrench from a crowbar" found bargains in every yard.[57] "You save, as a rule, *at least 66 per cent* of new part prices," according to one account from 1956, which went on to explain that "in the case of items like used hoods and doors, you get an extra bonus: paint, hardware, etc., are all included at no additional charge." Costlier new replacements, on the other hand, typically required several hours' labor to transfer odds and ends like hinges, trim, and in the case of doors, their glass and internal window regulators.[58]

Savings weren't the only reason to consider salvage yards instead of new-parts counters. Broad availability was another. "You can't very well go to your dealer and ask for a 1937 Cord fender or a 1941 La Salle grille," as one contemporary pointed out in 1956. "The wrecker is, in these cases, the last resort. If these parts are to be found at all, they will be found there. And this advantage is not confined to vintage models." As another explained in 1957, overhead costs and simple market risk prevented new-parts houses from maintaining inventories of every single part for every single later-model car: "New parts are expensive, and represent a substantial investment which, if not moved quickly, amounts to a slow return, and in some cases a loss. In contrast, however, the dealer in used parts has invested minutely. He can afford to hold on to a wreck until it is practically stripped clean and ready for scrap. Another thing, he has from a half to two acres upon which to keep all the tiny components, dubbed as 'slow-moving' by dealers' parts departments. This makes him an almost sure bet when it comes to small, scarce inessentials."[59]

In time this would become one of the most vital reasons why car enthusiasts of all stripes came to depend on and in many cases venerate the open-yard approach. But for most mainstream customers, the potential savings on doors, gauges, switches, tires, and other common parts were paramount. Bargains were also available on engines, transmissions, and other mechanical parts, but few contemporaries recommended buying salvaged bearings, brake components, or other safety-related components. "No saving is worth a risk," the do-it-yourself magazine *Auto Mechanics* explained in 1957, but "if you exercise good judgment in your purchase, and examine the merchandise carefully, chances are you'll get what you're after, in reasonable condition, and at a considerable saving."[60] A considerable saving, yes, but not rock-bottom: the owners and operators of these yards were by no means fools, and they did not take kindly to wrench-wielding visitors who tried to haggle the price of certain parts down below their market value. "We generally know the new and used prices of every part we salvage and sell," explained one in 1955. "We also know the market; we have to, it's our business."[61] Their operations may have looked a lot messier than those of the 1920s, but they were no less well run—differently run, yes, but no less well run.

In the 1960s and 1970s, two important sets of regulatory challenges at the federal level gradually began to alter how these yards did business. State, local, and federal regulation of the salvage trade was certainly nothing new. Several cities and states had long used zoning and licensing measures to prevent the uncontrolled growth and proliferation of urban yards, an effort that, combined with salvage yards' increasing need for outdoor space, contributed to a gradual shift within the business from cities themselves to their peripheries. About a dozen states had long required salvage titles on all junked cars, too, in some cases to prevent them from ever returning to the streets at all, in others to prevent them from being repaired and resold without clear notice that they once were junked.[62] The temporary intervention of the federal authorities into the business of dismantling and scrapping cars during World War II and the Korean War warrants mention here as well. But the new initiatives of the 1960s and 1970s greatly differed in scale and scope.

Consider the first: beautification. By the middle of the 1960s a number of Americans had grown weary of the blight of massive junkyards sprawling in every direction on the fringes of their cities. This, coupled with a desire to rein in both the well-publicized abandoned-car problem in urban areas and the proliferation of unsightly roadside billboards, led to the passage of an important piece of federal legislation championed by Lady Bird Johnson, the Highway Beautification Act of 1965. Chapter 5 delves into the many direct and indirect consequences of this act for the salvage trade, the used-parts business, and the automotive hobby. For now it will suffice to note that in the years that followed its enactment, billboards and litter were reined in, many junkyards erected solid fences or other barriers to screen their operations from view, and at least one new tool, the automobile shredder, made scrap recovery lucrative even when prices were low. Pioneered in the 1930s but more fully developed in the 1950s and 1960s, shredders allowed an entire junked car to be cut up and its materials separated for scrap automatically, without the need for labor-intensive manual disassembly and sorting. This aided the beautification effort by reducing junkyard inventories (and their associated sprawling blight), and it also helped curtail the abandoned-car problem by making the business of towing away stripped and burned-out hulks a reliably profitable endeavor.[63]

In time, however, the luster associated with the Highway Beautification Act faded. Ten years after it was signed, only about 10 percent of salvage businesses had built solid fences (or planted hedgerows) around their properties, and at the end of the 1970s more than two-thirds of the offending billboards targeted in the 1965 act remained in place. The problem was that the Highway Beautification Act, like many federal laws, relied on the states for enforcement. While some were quick to comply, others—New Jersey and Missouri, for example—posted 0 percent compliance rates in the late 1970s, while Georgia's rate was technically in the red because its officials allowed several hundred new and nonconforming billboards to go up. Federal funding was another problem. When the act was new, it received more than $75 million annually, but by the end of the 1970s that figure had dropped to $13 million, and further cuts loomed for fiscal 1980. With federal and state support spotty, the legislation was failing to live up to its promise, and garden-clubbers and environmentalists actively called for its replacement with a more consistently funded and evenly enforced measure.[64]

That never happened. Instead, Congress set in motion a second set of regulatory initiatives directed at the salvage industry by strengthening the federal government's regulatory power over solid-waste disposal sites. It did so first by requiring the Environmental Protection Agency (EPA) to study solid-waste toxicity shortly after the agency's creation in 1970, then by way of the Resource Conservation and Recovery Act of 1976, and finally with the Comprehensive Environmental Response, Compensation, and Liability Act of 1980, which created the toxic-cleanup "Superfund."[65] Regulations on groundwater contamination and toxic-materials recovery also proliferated during the 1970s. These new rules prompted some salvage firms to adopt gravel surfaces, others to pave their lots, and many with yards on dirt or gravel to adopt systematic procedures for the removal of each wreck's fluids and batteries upon arrival.[66] But other yards did little to clean up their acts, and from the 1980s to the present, state authorities coast to coast have worked with the EPA to identify those salvage businesses most egregiously out of compliance. Many have closed for good, and more than a few of their owners have been prosecuted and heavily fined.[67]

Other headline-making cleanup cases of the 1970s, 1980s, and 1990s instead involved the world of organized crime: drug rings that disposed of potential witnesses in the oblivion of wrecking yards, criminal organizations that operated out of junkyard office trailers, otherwise legitimate salvage operations that paid cash for stolen cars and parts, and other nefarious activities.[68] In short, although federal beautification-related shielding proceeded slowly during the 1970s, efforts to clean up salvage businesses on other environmental and legal grounds gained momentum.

Junking the "Junk" in Junkyards

Many salvage yards began to clean up their operations in other ways as well. Working against their industry's collective reputation for filth, crime, and especially disorder, they rethought, reorganized, and in many cases utterly transformed their businesses by mixing certain practices once common in the trade with several more recent innovations. Sometimes these shifts developed in response to externalities, as with the aforementioned advent of paved lots and fluid-removal procedures. Far more often, though, they merely meant that certain operators were no longer satisfied with the laid-back, open-yard approach to the business. The result, during the 1970s, 1980s, and 1990s, was the rebirth of the streamlined salvage model, albeit with some modern twists. In addition to those imposed by local, state, and federal authorities—salvage titles, visual shielding requirements, and air-, water-, and soil-contamination guidelines—five trends gave shape to this broader, but by no means universal, movement.

The first involved a revolution in the use of information technology. Salvage operators had always used the telephone to locate parts they did not have on hand to fill specific orders, and over time long-distance phone trees gradually emerged to facilitate this essential retail-counter operation.[69] During the 1970s salvage yards in some areas—metropolitan New York, for example—began to link up via teletype as well.[70] But personal computers enabled far more sweeping changes in the way that salvage yards did business. At first they were simply used to better manage wrecked-car inventories. "You have to have the ability to make the quick connection between what someone needs and what you have," explained Russell McKinnon of the Automotive Dismantlers and Recyclers of America in 1983. "Before computers, you needed a counter man with a great memory."[71] Some yards had counter men who were exceptionally gifted in this manner,[72] but most did not. Moreover, "because the same part may be used in a number of models"—and, on occasion, more than one brand of car—"software developed for the industry [enabled] cross-checks for these interchangeable parts."[73] By the end of the 1980s computers with modems enabled salvage-yard operators on the cutting edge to connect with one another and directly search each other's inventories, hastening the process of locating elusive parts.[74] Within another ten years some began to place their inventories on the Internet as well so that their customers could search for what they needed on their own, and digital clearinghouses soon emerged— AutoParts.com, Carparts.com, and so forth.[75] There is, however, an old saw among computer programmers regarding the reliability of data generated through search algorithms: "garbage in, garbage out." In other words, the results a database spits out when an operator enters a search function will be precise only to the extent that the original source data were recorded accurately in the first place. Thus, those that made best use of the possibilities inherent in this information revolution were those with systems in place to carefully and fully log each wreck and what it had to offer on arrival.[76]

The second trend involved the increasing ubiquity of portable and stationary crushers. Not to be confused with automobile shredders, tools of the scrap-metal trade that reduce entire cars into gnarled chunks of furnace-ready metal, crushers simply compacted them—typically vertically, but sometimes in all three dimensions to create "car cubes"—for easier storage and handling. Developed in 1967 and common by the 1980s, crushers closed a critical gap in the showroom-to-driveway-to-junkyard-to-scrap-yard cycle of automotive life by making it easier for salvage yards to boost their rates of inventory turnover. Once a car had yielded all of its most profitable parts, that is, a crusher made it far more economical for a salvage yard to haul it to the scrap yard. It was a simple matter of physics: eighteen to twenty-one crushed cars could easily and safely be stacked on a single flatbed tractor-trailer setup (fig. 2) versus but a handful if they were more crudely compacted by sledgehammers and backhoes.[77] Crushers also gave us some of television's most memorable scenes, from Michael Knight and his car KITT being dropped into one in season 2 of *Knight Rider* to the obliteration of Walter White and Jesse Pinkman's "Krystal Ship" RV in a crusher during season 3 of *Breaking Bad*.

Figure 2. Stacks of crushed cars, fresh from a salvage-yard crusher, arriving via flatbed trailer at a scrap-metal yard in Flowood, Mississippi, in 1972. (Photograph by Bill Shrout for the EPA's Documerica-1 Exhibition. Reproduced courtesy National Archives, Still Picture Records Division, Special Media Services Division, 412-DA-10526.)

The third trend centered on the rising social, political, and economic cost of land itself and of salvage-business licenses. Because of the sky-high cost of real estate, for example, businesses in Southern California in particular found over time that it had become prohibitively expensive to expand their operations horizontally. Some therefore adopted a more streamlined model, dismantling their wrecks entirely as they arrived and storing their parts. Such was the case at Coast G.M. Salvage, a one-and-a-half-acre yard in Long Beach where incoming cars were either stripped of marketable parts for storage and resale or, in the case of more valuable or rebuildable vehicles, stored intact on multilevel outdoor storage racks. The same was largely true at Jim Somsak's Atwood Recyclers and Auto Parts in Placentia. Bill Papke, on the other hand, kept at most a couple of wrecks at any given moment at an off-site, twelve-hundred-square-foot storage plot before dismantling them at his used-parts shop in Huntington Beach.[78]

Elsewhere, county boards and town councils tightened their salvage-business licensing requirements so that annual renewals became less and less routine, even for long-established yards. More important, their new rules and regulations often meant that new licenses for would-be entrants into the trade became virtually unobtainable.[79] This greatly enhanced the value of a grandfathered salvage license in many areas, so much so that marginally profitable open yards were sometimes bought by entrepreneurs who then looked to make the most of their precious new licenses by adopting a streamlined approach.[80]

The rise of corporate ownership, the fourth trend, had a similar result. Several chains of corporately owned yards emerged in the 1990s and 2000s, including Copart of San Mateo, LKQ of Chicago, Pull-A-Part of Atlanta, and Pick-n-Pull of Sacramento. Most of these chains retained the open-yard, you-pull-it business model rather than adopting the dismantle-and-store approach. But because they featured fully computerized inventories, neatly (and safely) arrayed rows of cars, very rapid turnover, and standardized, no-haggle pricing, they had a lot more in common with the fully streamlined yards of the 1920s than they did with the open graveyards of the 1950s and 1960s. Consider Copart. Established in Vallejo, California, in 1982, the firm expanded dramatically in the 1990s when its founder, Willis J. Johnson, teamed up with Barry Rosenstein, a partner in a venture capital firm in nearby San Francisco. The resulting infusion of cash allowed what was a small, three-yard operation in 1993 to develop into a ninety-one-yard chain of streamlined enterprises within a decade. Copart also broadened its reach by expanding into the once nebulous space between the insurance and salvage industries, serving as a middleman by buying totaled cars from insurers and auctioning them off to other yards. By 2002, Copart controlled 30 percent of this lucrative intermediate market.[81]

Because large-scale salvage operations like Copart were corporately owned, they were able to deal more effectively than others with the fifth and final trend of note: skyrocketing liability-insurance rates, particularly in the 1980s. This forced some to abandon the you-pull-it model altogether and others to better order their yards to make them safer for patrons and employees alike. "The well-publicized 'liability crisis' in this country has finally discovered the salvage yard operation. Insurance rates for liability alone have doubled, tripled and more for some operators," explained long-time salvage-yard chronicler Bob Stevens of *Cars and Parts* in 1989. "We know of several yards which have been forced to curtail public access . . . because of insurance requirements," he continued, adding that "we even know of one yard that has closed its doors because of the problem."[82] Larger chains like Pick-n-Pull and Pull-A-Part were better able to absorb these costs than smaller, more traditional mom-and-pop-style yards.

The salvage business has come a long way since the sprawling free-for-all of the mid-twentieth century. Although many open yards of the 1950s and 1960s variety persist, the overarching trend since then has been back toward the dominant model of the 1910s and 1920s, streamlined salvage. For ordinary do-it-yourselfers, this has certainly been a blessing, for by the end of the 1990s one no longer needed muck boots, coveralls, or even a portable toolbox to obtain an inexpensive part for an ailing daily driver. Indeed, because most used parts were by then a click away online, one no longer needed to visit any salvage yards at all.

But what of those who weren't simply looking to repair their grocery-getters economically? What of those who actually found inspiration, even beauty, in the rust and weeds of open yards? What of those who, like Bob Stevens, openly lamented the decline of these "vital resource[s]" and called on others to do their part to save them from extinction?[83] What, that is, of car enthusiasts?

The remainder of this book considers their experience.

2

Parts, Parts Cars, and Car Enthusiasts

The Art and Practice of Direct Recycling

Since the end of the 1960s, the automotive circle of life has been relatively straightforward for most cars, most of the time. Whether produced in a domestic factory or abroad, their lives began with journeys to dealerships by sea, rail, and/or truck—sometimes all three. A handful never made it. More than twenty-eight hundred brand-new Saabs, Volvos, and BMWs were utterly ruined in 2002 when the *Tricolor*, the Norwegian cargo ship carrying them, sank in the English Channel. When raised in 2003 by a Dutch salvage company, both the ship and its cargo were written off as total losses. Similarly, when the *Cougar Ace* listed badly in the North Pacific in 2006, the more than forty-seven hundred Mazdas on board were sold upon recovery as scrap rather than as cars. Though most of them were unharmed in the incident, Mazda's engineers and lawyers determined that the associated losses were preferable to the liability exposure that the firm would face were the cars instead sold new or used, or even broken up for parts.[1]

Of course, the majority of new cars manufactured over the years survived their journeys from the factory unscathed and ended up on dealer lots. Once sold, they led varied lives for an average of five to ten years, often in the hands of several owners, before most wound up in salvage yards.[2] Some arrived there sooner than others as the result of traffic accidents, premature mechanical meltdowns, natural disasters, fires, or, especially in northern cities, rust. Others stumbled or were towed in after many years and hundreds of thousands of miles of service. As in the late 1890s and early 1900s, a few instead were tucked away in backyards, barns, and other hideaways, some of which became the stuff of legend when they were discovered years or even decades later by collectors and other hobbyists.[3] But for most, salvage yards were the next logical step. There they rested until the owners of the yards determined that they had yielded all of the parts they were ever likely to profitably surrender. This took anywhere from a couple of weeks to several months, or sometimes years. In 1985, for example, *Smithsonian*'s James R. Chiles documented via a series of photos the process by which a Datsun pickup was transformed from a rough but complete vehicle to a picked-clean shell in less than ten days.[4] At the other extreme, *Cars and Parts* ran hundreds of feature articles on salvage yards in the 1980s, 1990s, and 2000s, many of which left their vehicles in place so long that trees grew through them.[5] Since the 1970s and 1980s, however, with access to compact crushers widespread, most cars have spent a lot less time than this in salvage yards before they were deemed ready to be sold as scrap. Once crushed, they were loaded in several stacks of seven or eight cars each on flatbed trucks and hauled to scrap yards. Scrap-yard operators then processed them, typically with automobile shredders, before selling what remained of them wholesale to steel companies in the United States and abroad. There they met their final fate in large furnaces that smelted them into steel—steel that was used, among other things, to manufacture new cars.[6] And so the cycle began anew.

From the factory to the showroom, from the showroom to the driveway, from the driveway to the salvage yard, from the salvage yard to the scrap yard, from the scrap yard to the steel industry, and from the steel industry back to the automobile factory: as one reporter summed up the process in 1972, "There may be a lot of recycled Detroit in that Japanese import."[7] This, indeed, is what most of us mean when we speak of "recycling." We mean used-up paper that is processed into pulp and then back into paper. Or we mean discarded aluminum cans that are smelted down and reborn as new cans. Or we mean plastic bottles that are shredded and processed into artificial down or carpet fibers. Or we mean our old iPods and Droids, dutifully returned to the store when replaced so that their parts can be separated and processed back into raw materials. Or we mean old Chevrolets, VWs, and Toyotas that are shredded into scrap and ultimately turned into new Chevrolets, VWs, or Toyotas. Or, if we haven't given it much thought, we might simply mean that we "recycle" when we throw a can or a bottle into a blue bin labeled with a Möbius strip of arrows rather than an ordinary trash can. In short, when we speak of *recycling*, we tend to mean *material flows*.

A more direct form of recycling takes place at the salvage yard. Consider those parts pulled from the Datsun pickup Chiles described in 1985, which ended up directly back in use in other Datsuns: wiper regulators, fenders, doors, suspension parts, and so forth. On the other hand, the picked-clean shell of that same truck might well have ended up in other pickups, too, but only indirectly—only after being crushed, shredded, processed into steel, stamped into panels, and assembled into brand-new trucks. *Indirect* recycling has long been the primary concern of scrap yards and other players in the secondary-materials market, from the days of the peddler and his cart to the era of the automobile shredder.[8] *Direct* recycling has instead long been the focus of automobile salvage yards, and the chief aim of those who frequent them, whether a car-insurance company looking to save a few bucks on a minor claim, an auto shop or a do-it-yourselfer seeking to complete a repair with inexpensive but high-quality used parts, or a gearhead in pursuit of scarce components for the restoration of an older car.

At times, however, salvaged parts and cars are neither indirectly recycled nor directly put back into use but are instead repurposed in creative ways—"folk recycling," as some have called it.[9] Artists have long used wrecked cars and wrecked-car parts in sculptures, public installations, and other pieces. State and local authorities have also used them to bolster barrier islands and to build up artificial reefs.[10] For their part, car enthusiasts have long engaged in bricolage with what they find in salvage yards, creatively repurposing, reshaping, and combining disparate parts and systems to build breathtaking hot rods, customs, and other one-off vehicles. The present chapter delves into this oft-neglected world of reuse and creative adaptation, beginning with a brief survey of artistic and utilitarian creations made with discarded cars and car parts before turning to five specific groups of automobile enthusiasts—hot rodders, restoration hobbyists, import and sports-car fans, car customizers, and street rodders—and their long experiences with direct recycling and bricolage.

Beauty and Utility

On old Route 66, approximately ten miles west of Amarillo, Texas, a row of half-buried Cadillacs covered with graffiti stands in silent tribute to the American luxury brand. Known as the *Cadillac Ranch*, the installation is the work of Ant Farm, a group of San Francisco artists—Chip Lord, Doug Michels, Curtis Schreier, and Hudson Marquez—who worked together on a number of projects in the 1960s and 1970s. The piece, which first went up in 1974 on the ranch of a wealthy Amarillo resident, Stanley Marsh III, consists of ten 1949–1964 model-year Cadillacs buried nose first, their trunks and tailfins exposed and angled toward the sky. With urban sprawl rapidly bringing Amarillo closer and closer to the *Ranch* each year, in 1997 Marsh decided to move the installation two miles farther to the west of the city. There it remains, the cars as colorful as ever.

Though Ant Farm's artists reportedly "felt a reverence for the stylish Cadillacs" of the 1950s and 1960s, their installation of them in the ground, nose first, does invite a pessimistic take, as if each of the jet-age cruisers were caught in the act of crashing back to earth. As a group, the apparently doomed cars were sometimes taken as an even broader statement. In 1983, Brock Yates, longtime *Car and Driver* contributor and outspoken critic of the American automobile industry of the 1960s and 1970s, used an image of the installation on the cover of his searing, Gibbon-esque indictment, *The Decline and Fall of the American Automobile Industry*. Nevertheless, the *Cadillac Ranch* was quickly embraced by enthusiasts and nonenthusiasts alike. *Smithsonian* took note of it in a lengthy article on the automotive afterlife in 1985. Both *Hot Rod* (in 1988) and *Cars and Parts* (in 2001) featured it in multipart stories about dream road trips along old Route 66. *Car and Driver* listed it among its ten best "cartifacts" in 1989 as well, and *Hot Rod* named it the "gearhead destination of the month" in June 2006. Part of the appeal of the *Cadillac Ranch* is that it has always been an *interactive* attraction: visitors are encouraged to graffiti the cars because the installation was intended to be a living rather than a static piece of art.[11]

Comparable installations popped up here and there in the 1980s. Six families teamed up with artist Jim Reinders in Alliance, Nebraska, in 1987 to create *Carhenge*, a tribute to the U.K.'s Stonehenge in which thirty-eight junked cars covered in gray paint stand in for the stones used in the original. After a brief spat with local and state authorities, who attempted to regulate the site as a junkyard, *Carhenge* went on to become a minor tourist attraction. Somewhat less imposing but perhaps more famous, thanks to its appearance in the 1992 comedy *Wayne's World*, was *The Spindle*. Built in Berwyn, Illinois, in 1989 by Los Angeles–based artist Dustin Shuler, *The Spindle* featured eight cars skewered on a large spike, a sort of automotive shish kebab. Another of Shuler's public installations, also in Berwyn, featured the exterior panels of a Ford Pinto flattened and mounted on an exterior brick wall, while another, *Death of an Era*, echoed the *Cadillac Ranch* by featuring a 1959 Cadillac on its side, punctured by a large nail, on a California State University campus. Also in California, Seattle-based artists Steve Badanes, Will Martin, Donna Walter, and Ross Whitehead built a sculpture called the *Fremont Troll* in 1991, a massive concrete rendering beneath a freeway overpass of a troll clutching a (real) Volkswagen Beetle. Outside the United States, a Prague artist placed a Trabant on a set of concrete elephant legs in the early 1990s; a French American sculptor created a fifty-nine-foot-tall pillar of concrete and crushed cars called *Long-Term Parking* near Paris in 1983; a Spanish artist encased several cars nose or tail first in a concrete block at an intersection in Jeddah, Saudi Arabia, in the early 1980s; and in 1986 the Canadian artist Bill Lishman beat Jim Reinders to the punch by a year by building *Autohenge*, his own forty-crushed-car tribute to England's distant past, outside Toronto.[12]

With the exception of those that were crushed, few of the cars in these large-scale installations were modified in any meaningful ways. They were repurposed, of course, but they were not transformed: the Beetle in *Fremont Troll* was still a Beetle, the Cadillacs in *Cadillac Ranch* still Cadillacs, and even Shuler's flattened Pinto was still recognizably a Pinto. On the other hand, consider those unwanted cars whose parts ended up in the work of James Corbett. A Brisbane, Australia, salvage-yard operator and aspiring artist, Corbett began to use parts from his wrecks to create unique sculptures in 1998. Within two years, his works of repurposed odds and ends were selling well enough that he was able to sell his salvage yard and devote himself to his art full time. Corbett's eerily lifelike sculptures—an orangutan with a body made from a DeSoto grille, a sheep with a "woolly" coat fabricated from hundreds of white spark plugs, a pair of kangaroos fashioned from air-cooled Volkswagen parts—were the result of a form of artistic bricolage known as assemblage, in which found objects are mixed together into new creations. Thus the junked cars used in Corbett's work were no longer cars per se but had instead become marsupials, apes, cowboys, and even a pair of daredevils in a scaled-down three-wheeled Morgan. Similar automotive assemblage featured in the work of John Kearney, who used car bumpers to create sculptures of animals in the 1960s and 1970s; Martin Heukeshoven, who used "naturally eroded surfaces" and "worn materials" from salvage yards and other secondary-materials depots to create naturally patinated scale-model cars; and Betty Conn and Edward X.

Tuttle, who used Kaiser hoods and fenders to create an iconic sculpture of Paul Bunyan in northern Michigan in the early 1960s.[13]

Others have created works of art from junked cars and car parts by seeking neither to repurpose nor to transform them but by discovering and exposing the beauty in these objects as found. In other words, they sought the beauty in junked cars *as junked cars*, rather than as found materials waiting to be fashioned or assembled into something new. Consider the work of Steve Moore, a photographer from Edmonds, Washington, whose work was featured in the March 1973 issue of *Car and Driver*. "What happens to old cars that are neglected is pretty clear," the feature began. "The forces of nature begin to work on them. . . . Pretty soon, what's left is a big chunk of metal that's been carved away by nature. It's no longer a car. It is a giant sculpture that usually ends up parked behind some gas station, slowly turning into iron oxide." This is where Moore stepped in with his camera, seeking to capture in these decomposing vehicles "an example of the continual battle between the positive creation of man and the negative destruction of nature."[14] Others working in the early to mid-1970s also sought to capture this tension, either in photographs or in paintings. But what matters for our present purposes is simply that artists like Steve Moore were out there, seeking beauty in in situ rust itself rather than in its potential as assemblage art.

Tom Merkel, on the other hand, was less interested in capturing moments of automotive decay than he was in the long-term processes that gradually destroy a car—the oxidation of paint, the rusting of metal, the yellowing of glass, the petrification of rubber, the rotting of fabric. He was also keenly interested in the lives that those cars lived. "To me they're like the pages from hundreds of family albums," he later explained, and as a result he felt compelled to save them from whatever fates awaited them at the ends of their road-going lives. He began to do so in the early 1970s, rescuing some from their final owners, others from used-car auctions, and more than a few from salvage yards. He had no intention of using or restoring any of them. Instead he simply arrayed them on the streets and later on several leased lots in Santa Barbara, California, as a sort of living work in progress about gradual, end-of-life decay. In time the authorities grew weary of his rather public collection and ordered him to eliminate the growing eyesore. He did so in 1984 by moving his cars to a private, eighty-acre tract of rural land far from civilization. There he had plenty of room to add to his collection, dubbed the *Car Garden*, which soon encompassed more than twelve hundred vehicles of every imaginable make and model. There he also had the peace and quiet necessary to enjoy them as they gradually returned to the earth. "Every one of the Car Garden's offerings was lived in—loved when new, taken for granted the majority of its life, and then discarded," *Car and Driver*'s John Pearley Huffman explained in 2005. "They have no new life in Tom Merkel's Car Garden," he continued, "but the lives of their owners, families, and drivers can be found in these artifacts."[15] Junkyard-sleuthing enthusiasts of the 1970s, 1980s, and beyond would have strongly disagreed with Merkel's decision to intentionally allow so many restorable classics to wither and rust. But they certainly would have understood, as *Cars and Parts*' Bob Stevens put it when touring a junkyard back in 1982, the desire to envision "what [an] old DeSoto looked like when it was new, [or] how much fun the original owners of [a] Ford ragtop had cruising through town with the top down."[16]

For every unwanted car that made it into a photograph, painting, or collection celebrating the beauty of in situ junk, however, and for every vehicle dismantled and remixed in assemblage art or repurposed intact in larger installations, tens of thousands of others have been creatively repurposed or rebuilt for second lives chiefly centered on utility and service. During the 1910s, 1920s, and 1930s, engines and other power-train components from worn-out or otherwise irreparably damaged cars often went on to power sawmills, pumps, boats, tractors, and even service locomotives. Sometimes an old chassis was reborn as an agricultural trailer, or an old drive shaft was hammered into a heavy crowbar. Other cars were recommissioned as rolling advertisements, while more than a few old roadsters and phaetons were converted into trucks.[17] The same remained true after World War II, when derelict examples of every imaginable make and model of car carried on as agricultural equipment, hunting stands, barbeque grills, furniture, and stationary advertisements.[18]

Partly because of their unique proportions, but mostly owing to their ubiquity, air-cooled Volkswagens were a popular choice. A number of Beetles and Microbuses ended up on poles as business advertisements in the 1960s, 1970s, and 1980s, while dozens of Beetles were sawed in half for use as kitschy displays in Gadzooks stores in the 1990s. At least one rear-end clip from a Microbus lived on in a shop as a bead-blasting hood, while another intact but thoroughly thrashed example was converted into a chicken coop (a common fate for cars of every make and model). One wrecked Beetle yielded its bent hood for use—after a bit of welding and the installation of an old cart wheel—as a makeshift wheelbarrow, while another gave a section of its roof for use as a tractor sunshade. Microbus shells were occasionally grafted onto the roofs of converted school-bus motor homes as second-floor additions, clips from totaled Karmann Ghias and Squarebacks sometimes became trailers, and at least one rusty Beetle lived on as an ill-advised tree house on stilts.[19] The possibilities were endless.

Seemingly endless as well, especially in the 1950s, 1960s, and early 1970s, was the supply of unwanted cars and parts. Used cars were abundant and cheap, so much so that the urban plague of the abandoned vehicle returned with a vengeance.[20] Salvage yards were supersaturated with wrecks, and neither beautification nor the technological sea change of the crusher and the shredder had done much as yet to shrink their inventories. For better or for worse, the thirty to thirty-five years immediately following the end of World War II was a time of unparalleled secondhand abundance in the automotive world. Not coincidentally, it was also a time in which a number of automotive hobbies—chief among them creating hot rods and customs, engaging in old-car restoration, working on imports and sports cars, and building street rods—became the genuine mass phenomena they would remain well into the twenty-first century.

Hot Rods

Although the term "hot rod" itself would not be coined until the middle of the 1940s, and although the practice it described would not become a broadly popular activity until the 1950s, hot rodding was born in the 1910s and 1920s. During those formative decades of the automotive age, car enthusiasm blossomed as the young, old, rich, and poor alike marveled at the racing prowess of Barney Oldfield and Ralph DePalma and at the sleek lines of top-dollar Duesenbergs and Stutzes. Before long, young enthusiasts without the means to purchase high-end vehicles like these discovered that, with a bit of elbow grease, they could modify their ordinary cars to perform and even look like much more costly models. Soon modified Fords, Chevrolets, and Dodges could be found in towns across the country and at local racing venues. Those who lived in Southern California also had access to several expansive dry lakebeds in the Mojave Desert north of Los Angeles, where they began to gather for weekend top-speed trials in the 1920s and 1930s. These young men[21] were the first hot rodders, and their souped-up cars the first hot rods.[22]

Hot rodding, the act of modifying one's car for improved performance and aesthetics, could be accomplished in several ways. Some were mechanically gifted enough to perform the necessary tweaks themselves by rewinding their ignition coils, shaving a few thousandths from their cylinder heads, or fabricating freer-flowing exhaust systems. Others relied on "speed equipment," special aftermarket parts designed to bolt onto an otherwise ordinary engine and produce more power: high-lift camshafts, high-compression pistons, larger-diameter carburetors, and the like. This equipment was produced by a small number of firms in the Midwest, and it ranged in price from the moderate (for simple items, like reground camshafts) to the obscene (for more complex products, like double-overhead-camshaft conversions). Even at the higher end of the market, however, the costs associated with using speed equipment to modify a low-end car remained far below those of walking into a dealership and simply buying a faster car.[23]

But speed equipment was not always cheap or plentiful, nor was it ever available for every conceivable need. Moreover, by the middle of the 1930s a number of established midwestern speed equipment companies had gone out of business, while others had failed to produce parts for newer designs—Ford's popular V-8 engines, for example. Although this gave a handful of entrepreneurs in Southern California an opportunity to break into the business for themselves during the latter half of the 1930s, for many hot rodders it simply meant a dearth of relevant and affordable speed equipment for their dry-lakes cars. Likewise, even the most talented enthusiasts were rarely able to circumvent the lack of affordable bolt-on parts altogether by reworking their cars entirely with parts they modified themselves.[24] But there was another option, a middle ground between over-the-counter speed equipment and self-made modifications: used parts from more expensive or better-performing automobiles.

Not everyone who could afford a Packard or an Auburn was a capable driver, of course, and neither was everyone who built a hot rod. Often enough, their automotive misadventures resulted in the appearance in salvage yards of wrecks equipped with valuable high-performance gear. This meant that savvy hot rodders could examine higher-end models in wrecking yards and come away with new ideas for their own cars. Or they could acquire, for relatively little money, larger carburetors, lighter wheels, or even complete power-train assemblies from high-end wrecks or speed equipment from mangled hot rods. Sometimes these used components would bolt straight onto their own cars; sometimes they required a little ingenuity. Either way, the junkyard quickly became a vital source of ideas and parts.[25] It also became an important gathering place, as chapter 3 explores in more detail.

During World War II, automobile racing and hot rodding were largely set aside. But after the war, high-performance enthusiasts quickly returned to their passion, only now they had a bit more life experience under their belts. Some of this experience meshed well with their prewar junkyard-scrounging days. Those who fought in the war had certainly learned a thing or two about improvised solutions, and thanks to wartime rationing, so had those who spent time on the home front. Many acquired formal training as mechanics and welders during the war as well. In short, California enthusiasts of the late 1940s and early 1950s were better prepared than ever to modify their cars; as they did, newcomers joined in coast to coast. And as hot rodding spread, new twists on the practice of modifying ordinary cars for improved performance and aesthetics surfaced. Some postwar hot rodders still built V-8 roadsters largely as they would have in the 1930s, while others focused instead on all-weather closed cars or on building specialized vehicles for drag racing, dry-lakes racing, and other forms of organized motor sport. In time hot rodding would spawn other niches, too, including street machines in the 1960s and 1970s and import tuners in the 1980s and 1990s, as well as two of the groups considered in greater detail later in this chapter: those who built sleek, boulevard-cruising customs and those who built street rods, arguably the ultimate expression of performance-minded bricolage.[26] What matters for our present purposes, however, is that the junkyard would remain a crucial source of parts and inspiration across these many high-performance niches throughout the balance of the twentieth century and well into the twenty-first. Let us turn, then, to precisely how and what these gearheads did with what they salvaged.

First, a number of enthusiasts with access to well-stocked salvage yards pioneered a practice during the mid- to late 1930s that their descendants in the 1990s and 2000s would call "OEM plus." The idea was simple enough: OEM plus normally involved building a car that a given OEM (original equipment manufacturer—Ford, Chevrolet, and so forth) never actually built, but could have built, using parts from other models made by that same OEM. Ford never produced a Model A equipped with its iconic flathead V-8 engine, for example. Instead the Model A, built from 1927 to 1931, came with a forty-horsepower in-line four-cylinder engine. A number of hot rodders active in the 1930s owned stripped-down Model As, and although a well-modified example of the four-cylinder breed could easily outrun Ford's own flathead V-8 models, introduced in 1932, many wondered what those new V-8s could do if they were souped-up for hot-rod duty. New and used V-8 Fords were too expensive for most shoestring-budget rodders, but wrecked V-8 models were abundant in Southern California's many salvage yards. Thus was born the original A-V8, a stripped-down Model A equipped with a Ford V-8 pulled from the mortal remains of a later-model wreck.[27] Arguably the first true OEM plus combination, the all-Ford A-V8 foreshadowed a great deal of what would follow among creative hot rodders and other gearheads over the next seventy-five-odd years. One enthusiast in the 1990s built a 1971 Chevrolet Monte Carlo SS convertible—a model that never existed—using parts from a 1971 Monte Carlo SS hardtop and a 1971 Chevelle convertible. Another fitted his Volkswagen Corrado sports coupe with a torque-rich four-cylinder Volkswagen TDI diesel engine to create a one-of-a-kind TDI Corrado. Others fitted Honda Civics with high-revving engines pulled from wrecked sister-brand Acuras; equipped their Corrados with all-wheel-drive systems from other Volkswagen Group brands; shoehorned big-block Pontiac V-8s into tiny Pontiac Fiero sports cars; and crammed New Beetles with potent twenty-four-valve VR6 Volkswagen engines.[28]

OEM plus also involved the duplication, using factory parts, of expensive or limited-production models. Consider a later example, the Porsche 914. Known outside the United States as the Volkswagen-Porsche 914, this was a relatively inexpensive sports car sold from model years 1970 to early 1976. Powered by mildly warmed-over versions of Volkswagen's Type IV air-cooled engines, used from 1969 in that firm's larger 411/412 cars and from 1972 in its decidedly unsporty line of Buses, the 914 was often praised for its handling and its looks but rarely for its off-the-line performance. But then there was the 914-6, a special version of the 914 equipped with Porsche's powerful flat-six 911 engine. Fast in a straight line and quick in the corners, the 914-6 was an outstanding car, but it was also rather rare when new and to this day it remains pricey. As a result, a number of ordinary 914s have been converted into 914-6s over the years, a relatively straightforward operation involving the use of original parts from salvaged Porsches to create affordable replicas of the iconic model.[29] In the same way, six-cylinder Camaros could be turned into more desirable V-8 models, and run-of-the-mill Mustangs could become high-end Shelby replicas.[30] On a much more mundane level, bigger brakes from heavier cars like Vista Cruiser wagons could be swapped into other GM products, heavy-duty axles from Lincolns could be installed in high-performance project Fords, quicker-ratio steering boxes from Firebirds could be swapped into any of a number of other GM cars, and so forth.[31]

A second approach to the use of salvaged parts and systems among high-performance enthusiasts involved cross-brand bricolage. This tack ditched the pretense of brand loyalty inherent in OEM plus and instead combined components from different makes—and often, different eras—to create truly unique rides. During the 1950s, this typically involved the installation of a much more powerful and state-of-the-art overhead-valve V-8 salvaged from a later-model car either into a classic prewar coupe or roadster or into a newer vehicle originally equipped with a more pedestrian power plant. Southern California's Dean Moon, for example, a well-known speed equipment manufacturer, installed an overhead-valve V-8 from a wrecked 1951 Studebaker Commander into his 1934 Ford coupe in 1953. He did so, according to a lengthy feature in *Hot Rod*, after "seeing the light" and recognizing that he had already taken his Ford's antiquated flathead mill as far as it could go in his ongoing quest for more power.[32] Others in the early to mid-1950s dropped late-model Buick V-8s into postwar Mercurys, Oldsmobile V-8s into Model A Fords, and Chrysler V-8s into everything from Model Ts to postwar sedans.[33] The floodgates opened further after Chevrolet introduced its line of small-block V-8 engines for model year 1955. Cheap, versatile, and easy to work on, by the middle of the 1960s the small-block Chevy had become the engine swapper's power plant of choice. Small-block motors found their way into Model As, 1930s coupes and roadsters, postwar customs, and, by the 1970s and 1980s, even E-Type Jaguars, Porsche 914s, Ford Mustangs, and Volkswagen Beetles. Others dropped Porsche 911 motors into Volkswagen Vanagons, turbo Renault engines into Lotus Europas, and Corvair motors into Volkswagen Beetles, Karmann Ghias, and Microbuses.[34]

Even more straightforward cross-brand swaps, like the one Dean Moon performed in 1953, required ingenuity and a lot of shoehorning. After all, Studebaker's engineers did not design their Commander V-8 to fit into a prewar Ford, and those at Chevrolet certainly never envisioned that their small-block mill would end up in a lowly Beetle. Specialty firms built kits and engine bell-housing adapters to facilitate a number of cross-brand and cross-era combinations as early as the 1950s,[35] but in theory any swap was possible if the rodder were skilled enough and the dimensions were right. In this regard, salvage yards were vital as a source not just of engines but also of ideas. "How do you know exactly which steering gear you will need for that new deuce coupe if you've never had a chance to look at all the gearboxes available," *Popular Hot Rodding*'s Tex Smith wondered back in 1967, adding that this "is why it is so important to wander around through a wrecking yard occasionally, getting a first-hand look at Detroit's finest." He then elaborated: "A hot rodder let loose in a good sized junkyard is rather like a child's first visit to the toy store—everywhere he looks there is something exciting. Out of such visits, however, come a multitude of good hot rod building ideas, including such things as better brake system combinations, interesting engine swaps, better steering set-ups, and hundreds of other ideas."[36] This was no less true thirty-four years later, when *European Car*'s Kevin Clemens took his son on a salvage-hunting expedition to several yards in search of a set of seats to fit the one-off rally car he was building. It was also no less true in 2007, when enthusiast Bill Hammer visited a salvage yard in search of a rear suspension setup for his radical rear-wheel-drive, Audi-V-8-powered Volkswagen Rabbit project. (After talking with the owner of the yard and sizing up several possibilities, he decided to go with the rear subframe from a 1990 Mazda Miata.)[37]

Smith, Clemens, and Hammer each knew well that one can never know what might work or definitely won't until one has taken the time to ponder and measure the possibilities up close. This is why the salvage yard has always been so vital to the high-performance bricoleur, whether a dry-lakes hot rodder in the 1930s, a muscle-car enthusiast in the 1970s, or an import tuner in the 1990s. It is also why the salvage yard was so important to another group of gearheads that emerged during the postwar era of second-hand automotive abundance: custom-car enthusiasts.

Customs

Prior to World War II, hot rodding was a relatively limited activity involving the construction of souped-up, stripped-down roadsters for use both on the streets and in dry-lakes racing. After the war, however, hot rodding fragmented into a number of smaller niches as it gradually developed into a nationwide phenomenon. Some hot rodders decided to focus their energy on the track, building dedicated dragsters and dry-lakes streamliners. Others focused on the street instead, building prewar coupes and roadsters with beautiful paint jobs, comfortable interiors, and in a number of cases modern power plants pulled from late-model wrecks. During the early 1950s, a number of others decided to focus their creative energy on these late-model cars themselves. Rather than buying a 1950 Mercury for its flathead V-8, that is, or a 1953 Chrysler for its Hemi or a 1956 Chevy for its small-block engine, these enthusiasts instead bought these cars in order to transform them into unique customs.

Often known as "lead sleds" because they tended to be heavy cars built for cruising rather than all-out speed and because of the widespread use of lead filler in their construction, customs quickly became a popular niche complete with its own lingo, magazines, and specialty shops. Every custom was by definition unique, although several modifications were common enough that they ultimately became standard fare. These included "chopped tops," with rooflines lowered through the removal of several inches or more from their supporting pillars; "deck and nose" jobs, involving the removal of chrome and other trim from trunks and hoods; "shaved" doors, with handles and other trim removed and filled; and "frenched" or sunken headlights, taillights, license-plate frames, and antenna bosses. Many also endeavored to reshape the contours of their hoods, trunk lids, and fenders, as well as the shape and composition of their grilles. It was intricate work, and although a handful of gifted customizers became famous for their high-end makeovers, most of the lead sleds built in the 1950s were the work of dedicated do-it-yourselfers.[38]

Whether taken on by a professional or an amateur, 1950s custom-car projects often began as late-model wrecks. As *Speed Age*'s Bud Unger explained in a 1953 feature on a customized Kaiser, "When choosing a car for customizing, it is wise to remember a damaged automobile is much cheaper to start with than an undamaged one—especially when the top is to be chopped or major grille and body changes are to be made." In other words, the extensive bodywork involved in creating a lead sled meant that heavily damaged cars could actually be a good place to start, depending on the nature of the damage and one's plans for the car. In the case of the featured Kaiser, the builder wanted to create a unique convertible, so he began with a 1951 two-door coupe that had suffered extensive rollover damage in an accident. He then removed the damaged roof, fabricated a removable padded top, and finished the rest of the bodywork with frenched headlights, grille modifications, and other custom staples.[39] Sometimes the decision to begin with a damaged car was less a matter of choice than circumstance. When Dick Bozeman of Damascus, Oregon, was involved in a major collision in his 1950 Chevrolet, his insurance company totaled the car. Rather than surrender it for the insurance settlement, Bozeman sent the car to a local custom shop, where the damaged sections were completely restyled.[40]

Or, for those with access to more than one hopeless wreck, several cars of different makes and models could be merged into a single and utterly unique custom. This is what Abilene, Kansas, resident Bill Albott did in 1950, when he combined the chassis from a 1940 Ford, the body from a 1937 Ford, the grille from a 1939 Buick, bumpers and various bits of trim salvaged from 1930s Chrysler products, and interior appointments pulled from a late-model Mercury.[41] It is also what Noah Poteet of Pete's Body Shop in San Antonio, Texas, did in 1953 when he combined a "wrecked '49 Ford convertible, a '39 Lincoln coupe and various assorted stock parts and custom accessories" to create a one-of-a-kind "Lincoln-Ford Hybrid." Grafting the Lincoln coupe's roof onto the damaged Ford convertible wasn't easy, but, as Rod and Custom put it in a feature on the car in 1953, that was where "the fun started!" "With the help of a torch, prodigious amounts of lead and a great deal of ingenuity and work," the magazine explained, "the Lincoln top was gracefully faded into the Ford body so that the rain gutter lines flowed smoothly into the Ford fenders." Then, "in order for the stock convertible doors and windows to fit smoothly to the top without leaks, the original convertible weather stripping was reshaped to fit the Lincoln top." The car was finished with a hood and grille from a 1951 Ford and the rear window from a 1946 Ford coupe, both grafted seamlessly into place on the once-wrecked car.[42]

Another approach to building a custom in the 1950s involved starting off with a clean and undamaged vehicle and then modifying its appearance using parts from wrecking yards. When Robert R. Hovey of Alton, Illinois, decided to restyle his 1949 Mercury convertible in 1951, the car was straight and true, apart from a few paint chips and door dings. But Hovey had the custom itch, and he wanted something different for the front of the car. Following "six months of periodic trips to the junkyards" and careful "measuring, redesigning, and pricing dozens of grilles," Hovey decided to build his own, using two horizontal grille bars salvaged from wrecked 1951 Kaisers. Like the junkyard-scrounging hot rodders described earlier, Hovey didn't know exactly what he wanted for his car until he saw, pondered, and measured it in a salvage yard.[43] In a similar manner, Dave Mitchell, a muffler-shop owner in Pasadena, California, transformed a straight but worn-out 1928 Ford roadster into a drop-top pickup in 1950. "One of the most novel features about the car," Hot Rod's Tom Medley explained in a brief photographic spread about the truck, "is the well-compartment built into the back of the body to accommodate the top when it is folded down." This was achieved using body panels salvaged from wrecks—specifically, "by grafting the back section from a 1928 Studebaker sedan in place" of the Ford's original sheet metal.[44]

In more radical cases, unique panels with specific curves unlike those of corresponding parts from any other car could be welded together from the many curves available in wrecking yards: a fender could be reshaped using metal from a roof panel, or a hood could be reshaped using the curves from a trunk lid. "All the contours you could ever use for even the wildest custom are obtainable at any large junk yard," Rod and Custom's Orin Tramz explained in 1953, adding that the "best bet is [a] yard specializing in late model cars," whose bodies were far more graceful and rounded than the simpler vehicles of the prewar period.[45] In less radical cases, fender skirts could be fabricated from old door skins, or taillight bezels from one car could be grafted onto the fenders of another.[46] The possibilities were endless, assuming one had an eye for the right junkyard parts and a talent for real-world bricolage.[47]

Over time, the custom-car scene gradually faded. More precisely, the custom-car scene as it was in the early to mid-1950s slowly slipped from prominence. High-end customizers continued to turn out wild show cars through the 1960s and 1970s, and in the 1970s and 1980s, 1950s-style lead sleds enjoyed a brief resurgence associated with the broader pop-cultural nostalgia for the 1950s that was prevalent at the time—witness the cars in

Happy Days, *American Graffiti*, and the *Back to the Future* franchise. In addition, the custom-car scene of the 1950s gave rise to another form of modified motoring during the 1960s, 1970s, and 1980s, the lowrider culture of the American Southwest. In a broader sense, customization also lived on as a practice: the techniques pioneered by those who sliced and diced their Mercurys, Kaisers, and Lincolns in the 1950s remained common among hot rodders, import and sports-car enthusiasts, and street rodders in the 1960s, 1970s, and beyond.

Bill Ridgeway, for example, the manager of VW Restorations and Customs in Manassas, Virginia, used a tried-and-true technique when rebuilding a picked-clean wreck of a 1970 Karmann Ghia in the early 1990s: he chopped the roofline by seven and a half inches. Likewise, when Paul Hearon of New Brunswick, New Jersey, rescued a battered 1965 Volkswagen Notchback from a vacant lot in the late 1980s, he decided to give the front end of the car the custom treatment by removing the front bumper and shaving and filling the remaining valence. And when Jeff Koch found a 1965 Volkswagen Microbus in a field in Colorado that had suffered serious damage to its roof in a hail storm, he did what the builder of the aforementioned 1951 Kaiser did some forty years earlier: he turned the wrecked van into an open-top car. Frenched license plates, antennas, and taillights remained common tricks as well, as did shaved door handles and cross-brand sharing of trim, lights, wheels, and interior appointments.[48]

Others took a page from the hot rodder's playbook and built what amounted to OEM plus customs. Often this involved the use of parts from wrecked cars to turn less expensive, more pedestrian examples into replicas of scarcer models. When *Hot VWs'* Rich Kimball came into the possession of a thrashed 1960 hardtop Beetle in the late 1980s, for example, he wanted to rebuild the car for his wife. But she wanted a Beetle with the more desirable "ragtop" option: a large, cloth sunroof. As Kimball put it, the 1960 Beetle "had already been clipped"—repaired following a serious front-end collision, that is, by having a major cross-section of sheet metal, or clip, from a donor car welded in place—and was thus of little value even if restored. Kimball also "had a sunroof section off a wrecked car" in his cache of parts, so he figured, "why not?" And so it was that a derelict 1960 hardtop became a nicely finished ragtop. This was a common modification for Beetles, a simple OEM plus job that involved cutting a ragtop section from a wreck, sawing a large hole in the center of the recipient's roof, and welding the sunroof section into place.[49] Other Volkswagen enthusiasts used clips cut from rare split-window Beetles (produced through early 1953) or only slightly-less-scarce oval-window models (produced through 1957) to turn more common examples from the 1960s and 1970s into split- or oval-window clones.[50]

From time to time, enthusiasts also built OEM plus customs of a different sort. Like those who put together the aforementioned TDI Corrado and Monte Carlo SS convertible, these customizers used original parts cut from wrecked vehicles to create cars the OEMs never actually built, but could have. When John Jones and Brian Phelps of Kustom Coach Werks in Grand Junction, Colorado, took to reconstructing a 1958 Volkswagen double-door panel van in 2001, for example, they decided to do something the factory never did. Panel vans have no side windows apart from those in the front doors, and they never came equipped with the skylights and sunroofs fitted to deluxe 23- and 21-window models. After welding in a roof clip from a wrecked 21-window Microbus, however, Jones and Phelps owned an utterly unique double-door panel deluxe.[51] Similarly, when Mike Kerwin of Tucson, Arizona, rebuilt his 1963 Volkswagen crew-cab pickup, he grafted on the wraparound rear-corner windows from a hopeless 1959 23-window deluxe, as well as four skylight windows from another derelict deluxe and the cloth sunroof section from a 1957 Mercedes sedan, to create a model Volkswagen never built: a deluxe double cab. Although his use of the Mercedes clip meant that the resulting truck was not altogether OEM plus in the strictest sense, sliding cloth sunroofs are always present on deluxe buses with skylight windows, so the truck that Kerwin created still looked factory built. But of course, it wasn't.[52] This distinction did not matter to Kerwin, nor to Jones, Phelps, Kimball, or any of the countless others who for many years have applied their saws and welding gear to what they find in salvage yards in order to build cars to suit *their* tastes, rather than those of the engineers who originally designed them. But this distinction did matter a great deal to our next group of junkyard-scrounging enthusiasts, the restoration hobbyists for whom factory originality was of paramount importance.

Old-Car Restoration

From the 1890s through the 1920s, automobiles were seldom "restored." At most they were "refurbished," as those who owned or sought to sell older models worked to squeeze a few more years of service from them. Among owners, such refurbishing often entailed installing a newer ignition or lubrication system and perhaps repainting the car so that it would run and sparkle like new.[53] Among dealers, who began to struggle with the so-called used-car problem during the 1910s, refurbishment sometimes involved more radical measures to help sell an otherwise undesirable car: installing a sportier body to turn an old sedan into a stylish speedster, converting an old touring car into a serviceable truck, and the like.[54] "Restoration," on the other hand, referred to the process of returning a worn-out car to showroom-new condition, not because it would otherwise be unattractive or unserviceable, but instead because of its intrinsic monetary, sentimental, or historic value. Restoration therefore differed from refurbishment less in process than in motivation. Unless one older car was modified extensively while being *refurbished* for continued use, that is, the process of *restoring* a second older model would have looked much the same: a thorough cleaning, a tune up, a repaint, and perhaps some new upholstery. But while the former would have been refurbished because its advanced age made it otherwise undesirable, the latter would have been restored precisely because its advanced age actually made it desirable.

At the dawn of the twentieth century, of course, "advanced age" meant something very different when applied to automobiles than it would a few decades later. During the 1900s the automotive press often ran short features on cars that were deemed to be "ancient," "old-timers," or "veterans," but these were often as little as three model years old and rarely more than six.[55] The reason they were viewed as relics in spite of their relative youth was that in the 1890s and 1900s, the automobile itself was still new, and the fundamentals of automotive technology were evolving very rapidly. Thus a three-year-old car could indeed have been quite old, functionally and aesthetically. As a result, "ancient" cars were sometimes given new leases on life through mild or radical refurbishment by urban and suburban secondhand dealers, but more often they met different fates. A few were consigned to the scrap heap, although this was uncommon in the 1890s and early 1900s. Many found new owners in small towns and rural areas instead, where the obsession with up-to-the-minute styling common among urbanites held less sway. Others, especially those still owned by wealthy early adopters who had since moved on to more fashionable and technically sophisticated designs, were simply "tucked away in private stables" where they languished in a state of disuse.[56] Not until the 1910s would any of these junked, rebuilt, resold, or otherwise forgotten old cars be restored, even on a limited basis, and not until the early 1930s would their recovery and restoration begin to develop into a recognizable hobby.[57] Simply put, in the 1890s and 1900s, sufficient time had not yet passed for any of them to have become collectible, and thus restorable. As chapter 6 explores in greater depth, the passage of time alone—more precisely, an object's *age* alone—is not the only factor of importance in establishing collectability. But at least in the 1890s and 1900s, their lack of age alone sufficed to keep most ancient cars from being seen by anyone as worth collecting.

During the 1910s, however, elites on both sides of the pond began to pay attention to these older relics. In England, Sir David L. Salomons and the Duke of Teck sponsored a research program in 1911, undertaken by a Mr. Edmund Dangerfield, to track down and document early examples of road locomotion with the aim of establishing a permanent museum "in order to prevent further regrettable losses of cars and cycles which daily become of greater interest." The following year a temporary exhibit debuted, and two years later Dangerfield's full museum opened its doors in London.[58] In 1913, the organizers of a new-car exhibition in Pittsburgh offered a cash prize to the oldest car that could make it to the show under its own power. A 1900 Winton, "chug[ging] . . . with as much speed and almost as much power as she showed" when new, claimed the prize.[59] Three years later, Elwood Haynes of the Haynes Automobile Company launched an effort to locate the oldest car bearing his name that was still in active service. Following an exhaustive search that turned up a number of contenders, the victor, James E. Howard of Jeffersonville, Indiana, was presented with a brand-new 1917 Haynes Twelve in exchange for his contest-winning 1897 two-cylinder model.[60] For its part, the automotive press began to wax nostalgic about the early days of motoring in the 1910s and 1920s. *Motor Age* and *The Automobile* did so by printing retrospective looks at the 1890s through the lens of their long-running sister publication, *Horseless Age*.[61] *Motor, Motor Age,* and *The Automobile* all ran celebratory features on surviving "old-timers" from the 1890s and early 1900s as well. These included an 1891 Panhard owned by a French clergyman since 1894, featured in both the 1910s and 1920s; a ten-year-old Cadillac that did well for itself at a Royal Automobile Club shakedown in the U.K. in 1913; a Parisian Léon

Bollée, built in 1897, which was used across South America before coming to the United States in 1900, featured in 1913; and a homemade car built by Achille Philion of Chicago in 1893, celebrated in 1920.[62]

By the 1930s, antique restorations, while still less common than continuing-use refurbishments, had nevertheless become sufficiently widespread among middle-class Americans to support small-scale car shows and the founding of the Antique Automobile Club of America (AACA, 1935) and the Horseless Carriage Club of America (HCCA, 1937). Members of the AACA and the HCCA were interested above all else in what they called "antiques," typically vehicles from the 1890s–1910s (HCCA) or the 1890s–1920s (AACA).[63] After World War II, the old-car restoration hobby broadened to include "classics," defined as the luxurious and powerful makes of the 1920s–1930s: Duesenbergs, larger Packards, Auburns, and the like. The hobby continued to grow, and by the time it hit its stride as a broad-based activity in the late 1950s, a third class of cars had joined the avocation's ranks. "Special-interest autos," as this final class was known, included the mass-market cars of the 1920s–1930s (mostly Fords, Chevrolets, and Dodges).[64] Later, during the 1970s and 1980s, the special-interest class would expand to incorporate the mass-market cars of the postwar years as well.[65]

As the old-car restoration hobby broadened and matured in the 1950s, it began to receive regular coverage in broad-based periodicals like *Car Life*, *Motorsport*, *Auto Life*, and *Motor Trend*. It also spawned several new titles developed specifically for restoration enthusiasts, notably *Hemmings Motor News*, first published in 1954, and *Cars and Parts*, which debuted in 1957. Some of the earliest postwar coverage focused on the emerging upper end of the old-car hobby: private individuals who could afford to spend good money on a classic—typically $1,000 to $5,000, at a time when median household income hovered around $3,900—and who could also afford to pay a shop to restore and maintain it. Furthermore, according to the 1952 *Automotive Yearbook* of the men's magazine *True*, the average classic enthusiast was "very contemptuous of Detroit's stock cars today, a firm subscriber to the theory that 'they don't build them like that any more.' . . . Other cars don't move him. Antiques, hot rods, California customs, and his family car are beneath consideration."[66] There were exceptions, of course. Some, including the well-known racing driver Phil Hill, did their own work on their classics. Others restored antiques *and* classic cars, caring little for the class distinctions that ostensibly stood between a 1908 Buick and a 1934 Packard Twelve.[67] Also, for every pretentious collector of high-end classics, there were many more who restored antiques.

Antique enthusiasts tended to be a bit lower on the income scale, and they also tended to do their own work. This was the case back in the 1920s and 1930s, when the old-car hobby centered on antiques first emerged, and it was no less true in the 1950s. As Frank Cetin of the general-interest magazine *Cars* explained in 1953, "Most people who start out restoring antique automobiles begin with any car they happen to get hold of." Then, as they settled into the hobby, they tended to gravitate toward one or another make or one or another span of years. But what really mattered, Cetin continued, wasn't "how or why these citizens get interested in restoring old cars or what kind of antiques they prefer. . . . They are all in the hobby for the same reason—the pride and joy that comes from transforming a pile of junk into an authentic and usable automobile. Authentic to them, by the way, means authentic right down to the last lock-washer."[68]

Doing the work of accurately restoring an antique meant finding and procuring spare parts, and in the 1950s this was often quite a challenge. This was long before specialty firms would reproduce the sheet metal, trim, and mechanical parts required, and as *Rod and Custom* put it in an article describing the restoration of a 1910 Buick touring car in 1954, "You can't [just] run down to the nearest wrecking yard and buy a fender or two." Instead, "You must hire an experienced craftsman who can duplicate exactly a missing fender from an old photograph or a smudgy sketch."[69]

Wrecking yards were often of limited use to antique enthusiasts in the 1950s for several reasons. Sometimes the car in question was never produced in large numbers to begin with, making the survival of derelict examples in salvage yards unlikely at best. Sometimes wrecked examples with the parts one needed were indeed out there somewhere but had been exposed to the elements for so long by the 1950s that virtually all of their parts were useless. Sometimes their scarcity was more deliberate. As we have seen, manufacturers and dealers destroyed thousands of "old-timers" during the 1920s and 1930s in an effort to remove competing used cars from the market. And then there were the wars. Thousands of antiques were fed to the scrap-metal furnace during World War II and the Korean War, artificially magnifying an increasingly desperate shortage of antique cars and parts.[70]

For the restoration enthusiast, this left three options: hiring a specialist to fabricate the needed parts, as *Rod and Custom* suggested; finding other hobbyists with similar cars and swapping for what was needed; or locating and purchasing a "duplicate car," more commonly known as a "parts car": a wrecked or otherwise hopelessly worn-out vehicle used as a rolling source of spares.[71] The first of these options was out of the question for most budget-minded antique do-it-yourselfers, while the third grew less and less possible each year precisely because so many were joining the ranks of the hobby. This adversely affected not only the chances of locating a parts car for use as a donor, but also the odds of finding a suitable antique project to begin with. "Discouraging as it may be," *Car Life*'s George A. Parks lamented in 1954, "most collectors feel that the chances of finding a pre-1920 car not in the hands of an enthusiast are a thousand to one!"[72] Consequently, for antique enthusiasts in particular but also for other old-car hobbyists, the second option gradually gained favor. By the 1970s, the resulting large-scale swap meet had become a vital seasonal tradition.

Other restoration enthusiasts had better luck finding complete parts cars for their projects, and they also found it easier to locate odds and ends in salvage yards. This was especially true among the special-interest hobbyists who joined the old-car fold in the 1950s. Because they focused on the newer, mass-produced cars of the 1920s and 1930s—Model A Fords, for example, as opposed to the high-end classics of the same period and the antiques of an earlier age—they had little trouble sourcing what they needed. "Ordinarily, the A's are purchased in near junk condition; some are merely running on their nerve, while several have been lacking completely in the wheels department," explained a 1954 *Car Life* feature on an early Model A Ford club and its members. The parts required to breathe new life into these machines were plentiful, the piece went on to clarify, with some still available new from Ford dealers. *Motorsport* detailed another option for Model A enthusiasts, also in 1954: "Replacement transmissions, rear ends, body parts and miscellaneous hardware" for the venerable Ford "can be picked up in any junk-yard, usually in better than good condition." Restorable examples and parts cars were available for a song, and the same was true for those who rebuilt Plymouths, Dodges, and Chevrolets from the same era.[73]

During the 1970s and 1980s, the special-interest niche expanded to include everything from 1940s and 1950s Cadillacs and Chevrolets to 1960s Mustangs, GTOs, and even postwar imports like Volkswagens, MGs, Jaguars, and Porsches. As this happened, a robust aftermarket reproduction-parts business developed for many of these later cars, especially Mustangs, Volkswagens, and MGs. A bit later, several OEMs also established "classic" or "heritage" centers to support the preservation and continued use of their earlier models, as well as licensing programs to better regulate the reproduction-parts business. Though this was not entirely new—Aston Martin, for example, has long supported every model it has ever made through its Works Service Department, established in 1924—it did become a more widespread phenomenon in the 1990s and 2000s with the establishment, among others, of General Motors' Service Parts Operations in 1991; Ford's Restoration Parts Licensing Program in 1993; Mercedes-Benz's Classic Center in Germany, also in 1993; Volkswagen's Classic Parts Center in 1997; and Ferrari Classiche in 2006.[74]

At the same time, salvage yards remained vital—indeed, for those who owned a newer special-interest car, they came to be an even more abundant source of restoration projects, parts cars, and individual parts and systems than they had been for the earlier special-interest enthusiasts of the immediate postwar years. This was due in part to the fact that latter-day special-interest cars like 1955 Eldorados, 1960 356s, and 1965 Malibu SSs were simply newer and thus more likely to have parts-car counterparts in salvage yards than any prewar car. In addition, very few of these postwar models had ever been subjected to forced destruction by new-car dealers or a scrap-hungry military. Finally, low scrap prices through the 1950s and 1960s meant that salvage inventories of postwar models remained bloated throughout the period. Thus it was easy, as *Hot Rod* would report in 1978, to find used parts for early Mustangs in wrecking yards. It was also easy, as *Popular Hot Rodding* would note a couple of years later, to find an inexpensive wrecked Camaro and rebuild it using salvaged parts. And, as regular readers of *Cars and Parts* would find over the course of the 1980s and 1990s, wrecking yards with salvageable machines from the 1950s and 1960s abounded coast to coast.[75] Thus, although prewar cars continued to surface regularly during the 1960s, 1970s, and 1980s, few benefited quite as readily from the postwar era of secondhand abundance as did latter-day special-interest enthusiasts.

To further illustrate the point, let us turn to the postwar experience of another group that ultimately became part of the special-interest fold: import and sports-car fans.

Imports and Sports Cars

During the 1950s, small sedans and sports cars bearing European nameplates gradually began to appear on American streets. GIs returning from Cold War duty in Europe brought these Renaults, Alfa Romeos, Triumphs, Jaguars, Porsches, Volkswagens, and other foreign-made cars back in droves. By the middle of that decade, some were also available from specialist dealers located primarily, at least at first, in West Coast and northeastern cities. Thus, while most Americans, especially those in the vast heartland, bought large domestic V-8-powered cars during the Truman and Eisenhower years, a small but growing cadre on the coastal urban fringes went instead for smaller European models. Likewise, while most Americans with an interest in motor sports followed stock-car or hot-rod drag racing during the postwar years, the typical import and sports-car enthusiast paid much closer attention to European contests, especially Formula One and Le Mans, which emphasized endurance and handling as well as straight-line speed. By the middle of the 1950s, specialist periodicals like *Sports Cars Illustrated* and *Sports Car Graphic* had hit the shelves, and sports-car races at Watkins Glen, Laguna Seca, and other American venues were attracting thousands of fans each year.

American enthusiasm for European racing was not entirely new. It first emerged during the 1930s among the East Coast elite, whose frequent trips abroad exposed them to upper-class racing circuits in England, Germany, France, and elsewhere. Efforts among these socialites to establish similar circuits in the United States were interrupted by World War II, but as that conflict drew to a close, wealthy Bostonians founded the Sports Car Club of America (SCCA) in 1944. Though it was thoroughly elitist in character, the SCCA was also an association for amateurs. This meant that while its members might have aspired to elite standards of taste, especially in motorcars and styles of racing, all one really needed to get involved in SCCA competition was a European sports car. In short, what was new in the 1950s was the participatory nature of European-style racing in the United States.[76]

Of the many foreign nameplates one might have seen in Boston, New York, or Los Angeles during the 1950s, however, only Volkswagen established a thoroughly widespread network of dealers. By the middle of the 1960s Porsche, Fiat, Jaguar, and newcomers like Saab had respectable dealership networks as well, but during the 1950s they did not. Neither did Alvis, Jowett, Singer, Lancia, DKW, Simca, and many other brands, some of which were backed by only a single U.S. importer. In practical terms, this meant that service and replacement parts for these European vehicles were difficult to come by, which was problematic both for those who used their Renaults to commute to work and for those who raced their Jaguars at SCCA meets. The metric system was at least in part to blame for this situation, but the general scarcity of the cars themselves was the chief culprit. So it was that wrecked examples came to be a vital source of parts for these cars. This was true of everything from carburetors, fenders, and trim to more basic odds and ends, like metric nuts and bolts, which were no less difficult to come by in some parts of the country.[77] Indeed, especially during the 1950s, the dearth of other options often meant that salvage yards were substantially more important for these enthusiasts than they were for any other group—more than for hot rodders searching for bargain motors, more than for customizers hunting for a particular curve or grille, and certainly more than for ordinary do-it-yourselfers looking to save a few dollars on a replacement axle. The following chapter takes a closer look at how this situation contributed to the emergence of European-specialist salvage yards in the 1950s and early 1960s, especially in coastal areas. For now, consider instead two key ways in which the import and sports-car hobby changed in later years.

First, by the early 1970s a number of imported sedans, wagons, vans, and sports cars from the 1950s and 1960s had become collectible and had thus become restoration candidates. In other words, they had become part of the special-interest old-car hobby, and as with other kinds of special-interest cars, later and later models would be added to the fold with every passing year. This meant that by the 1980s and 1990s, enthusiasts were tearing apart and rebuilding worn-out imports and sports cars originally manufactured as late as the 1970s. And when they did, they often relied on salvage yards for major and minor parts and systems, just as other special-interest gearheads did. Unlike their 1950s predecessors in the import and sports car hobby, that is, these latter-day enthusiasts turned to salvage yards not to keep their obscure cars on the road when new, but to bring them back to life after many years of service. Some of those who undertook these projects shared the goal of other restoration hobbyists: period-correct authenticity, down to the last nut and bolt. Such was the case, for example, when Richard Fitschen of Tempe, Arizona, restored a 1960 23-window deluxe Volkswagen Microbus that he found in a storage lot in the late 1980s, using an unmolested dashboard clip and many other components from a parts bus. Others sought instead to restore their vintage imports with an eye toward daily use in modern traffic. When Scott McDuffie of Greensboro, North Carolina, rescued a 1972 BMW 2002 from a junkyard in the 1980s, for instance, he rebuilt the car for daily use with a modified suspension and a larger engine. More than a few others turned older sports cars into vintage racers as well. *European Car*'s Brett Johnson restored two Porsche 356s in the early 1990s, and during the course of this work he bought a basket-case 1951 356 for use as a parts car. After finishing his restoration projects, Johnson decided to turn the parts car into a vintage racer, since it was rare enough to save from the junkyard but too far gone for an all-out restoration.[78] Import and sports-car periodicals from the 1980s and 1990s brimmed with stories similar to these, as MGs, Porsches, Jaguars, and Volkswagens settled into their new roles as special-interest cars.

The second shift within the import and sports-car hobby during the 1960s, 1970s, and 1980s involved the use of parts from wrecked examples in radically different projects. E-Type and XJ Jaguars from the 1960s and 1970s often yielded their complete rear suspension assemblies for use in street-rod projects, as the following section on street rodding examines in greater detail. Likewise, components from worn-out imports often served as the basis for kit-car projects. The fiberglass-bodied kit car first emerged in the 1950s and 1960s as an inexpensive way to build a special-bodied vehicle, typically using the frame and running gear salvaged from a more pedestrian model. Especially during the 1950s and 1960s, some kit cars were sporty roadsters sold as cheaper alternatives to the genuine European sports cars of the day. Others were open-top dune buggies meant for off-road use and beachcombing. Later, during the 1970s and 1980s, other kit cars were produced that mimicked everything from prewar Bugattis to postwar MGs, while some were meant instead as inexpensive alternatives to more modern exotics.[79]

Sometimes Porsche, Jaguar, and MG running gear was used to power these replicas, and sometimes domestic engines and chassis components from Pintos, Mustangs, and V-8 Chevrolets were used instead.[80] But by far the most common donor for a kit-car project was the Volkswagen Beetle. The Devin "D," for example, a $1,495 fiberglass sports-car kit produced at an El Monte, California, factory, used the engine, transaxle, and wheels and the steering, braking, and suspension systems from a Beetle. "Local wrecking yards will usually yield a bodyless VW with enough parts for under $500," *Motor Trend* advised in a feature on the Devin in early 1960.[81] Similarly, EMPI of Riverside, California, introduced a fiberglass dune-buggy kit called the Sportster in 1964. Based on a shortened VW chassis included with the kit, the Sportster required a Beetle's gas tank, wheels, transaxle, engine, and front and rear suspension. After sourcing these parts from a wrecked VW, an enthusiast who bought and finished a Sportster kit could send EMPI the bare, standard-wheelbase chassis from their donor VW for credit toward the original purchase price. EMPI then modified these donor chassis for use in other Sportster kits.[82] Volkswagen engines and chassis were also used in numerous other dune-buggy kits, replica MG TFs and Porsche Speedsters, and one-off sports cars like the Bradley GT.[83]

Without well-stocked salvage yards and gearheads willing to tackle a challenge, far fewer imports and sports cars would have survived the ravages of time, and far fewer Beetles would have been reborn as beachcombers and svelte sports coupes. Likewise, without abundant wrecking yards and creative enthusiasts, street rodding, our final postwar niche, could never have emerged and thrived.

Street Rods

By the middle of the 1960s, a number of hot rodders were convinced that their hobby had lost its way. In place of the simple, cheap, and versatile performance-oriented roadsters of the 1930s, 1940s, and early 1950s, there were now a number of much more costly niches. Drag racing had evolved into a serious endeavor, even for amateurs. So had dry-lakes racing. Similarly, the need to keep top-dollar show cars pristine had sapped a lot of energy from the street. Finally, and for some most troublingly, the American automobile industry had begun to co-opt what was left of traditional hot rodding with their muscle cars and pony cars, seriously fast machines built using every go-fast trick in the old-school hot rodder's book. Unlike genuine hot rods, though, these mass-produced cars were available on easy terms, came with warranties, and did not require meticulous end-user planning, patience, and parts-scrounging. To top it off, many of these OEM "factory hot rods" were drag-strip-ready right out of the box. Thus, by the middle of the 1960s, a high-performance enthusiast could either spend a lot of money on a one-off dragster or a rarely driven show car, or he could send GM or Ford a check each month for a brand-new street/strip car. To the dismay of a number of gearheads—some of whom were old enough to remember the hot rods of the 1930s, 1940s, and 1950s, but most of whom were a bit younger and had only come of age in the 1960s—traditional, technically creative hot rodding seemed to have been marginalized.[84] Their collective response was the street rod, which emerged in the 1960s and matured in the 1970s.

In theory, street rodding was supposed to mimic hot rodding as it was in the 1930s, 1940s, and early 1950s. Broadly speaking, it did: it involved the construction of performance-modified roadsters and coupes and their frequent use on the street. But as is often the case among nostalgic reincarnations of earlier activities, street rodding differed from hot rodding in a number of critical ways. Street rodders built fast cars, for example, but they rarely raced them. They also used a myriad of automotive systems unavailable to their predecessors back in the "good old days," many of which had as much to do with cruising comfort as they did with speed. These included power steering, power-assisted brakes, stereo systems, air-conditioning, larger and more powerful V-8 engines, and even automatic transmissions, power windows, and power seats. Perhaps most important, street rodders spent a lot of time and energy on the way their cars looked. They fussed over their paint jobs and interior appointments—not nearly as much as the owners of one-hundred-point concours show cars, but certainly much more than earlier hot rodders.

But it's the way they went about building their cars that concerns us here, and in this respect, street rodding closely resembled hot rodding. Street rodders began with prewar cars, often low-dollar, basket-case examples, and reconstructed them with high-performance and custom-oriented parts and systems. Many of their finishing touches came from aftermarket companies, but the basic components often came from wrecked late-model cars and were pulled together in a manner reminiscent of 1950s bricolage. Like those who once built custom cars from salvaged body panels and grilles, street rodders tended to live by the maxim that you never know exactly what you're looking for until you see it. *Street Scene*'s Joe Mayall summed it up well in 1981:

> I've been a wrecking yard hound for as long as I can remember, and some of my best times were when I would just wander around in one, tape measure in hand, looking into and under every vehicle that was there. Sometimes I had a specific need, but often I used the opportunity to check out things that I didn't have a thought of using but might result in an idea for future use. You'd be surprised how much creative thinking can be generated by looking under a smashed up car. I'm sure it was a trip to an automobile dismantling yard that resulted in the current popularity of the Vega and Mustang steering in street rods.[85]

Exploring, measuring, and working out the possibilities in the mind's eye: this is how street rods were built, especially in the 1960s and 1970s.

Typical among the creative cross-brand hybrids that resulted from junkyard-scouting trips like these was a 1933 Ford panel truck featured in *Rod and Custom* in April of 1973. Built by Jerry Dawson, a drag racer from St. Louis, Missouri, the Ford was rebuilt with a steering system from a Chevrolet Vega, chosen because "it work[ed] well and it [was] the proper length" for the Ford's cab, and because it "contain[ed] all built-in wiring as well as turn signals, horn, ignition key and lock." The all-inclusive Vega column, *Rod and Custom* noted, would be a "neat setup for other cars also." The panel truck was fitted with the rear suspension from a 1971 Jaguar as well. This was a popular choice among street rodders, since the entire rear suspension and differential assembly of E-Type and XJ Jaguars was a single modular unit that came out of a donor car as one assembly and could be bolted into another chassis with minimal fuss. Dawson finished off his panel truck with a late-model small-block Chevrolet V-8, Buick wire wheels, and disc brakes from a Camaro.[86] Also true to form was a 1936 Ford roadster built in 1967 using a late-model American Motors V-8 and a Chevrolet rear end, pictured in *Hot Rod* in May 1976.[87] An even closer approximation of the street-rod ideal appeared one year earlier in the same magazine: a 1918 Model T put together in the early 1970s with a steering system from a late-model Chevrolet Corvair, a differential from a 1969 Jaguar, a small-block V-8 engine and turbo 350 transmission from a 1972 Chevrolet, a gear selector from a Ford Mustang, a pair of front spindles from an MG, and a set of wire wheels from a Buick; the car was finished off with a custom-stitched interior and low-key speed equipment.[88] Other examples filled the stalls at street-rod meets from coast to coast, as well as the pages of *Rod and Custom, Hot Rod, Rod Action, Popular Hot Rodding,* and *Street Rodder.*

Things began to change as both the street-rod hobby and the average street-rod hobbyist matured in the 1970s and 1980s.[89] The most important shift involved the emergence of specialty firms catering specifically to those who built street rods. By supplying interior kits, wiring harnesses, various engine and chassis adapters, and other odds and ends, they made it less and less important for a would-be street rodder to visit a salvage yard and size up the possibilities. Fiberglass coupe and roadster bodies, first available in the 1960s and widespread by the 1970s, also allowed street rodders to bypass the preliminary process of obtaining a rebuildable prewar shell. By the end of the 1970s, it was even possible to build an entire street rod from scratch, using only reproduction parts.[90] Rebuildable prewar shells remained a holy grail of sorts among street rodders, and the quest for ever-scarcer genuine components for their projects contributed directly to the emergence of the treasure-hunting trope. Nevertheless, by the end of the 1970s the age of the fiberglass reproduction was here to stay.

Some lamented this. "The all glass and new, hi-tech cars are truly lovely imitations," Kathleen Piasecki of Antioch, Illinois, wrote in a letter to *Street Scene* at the end of 1980, "but in this writer's opinion they will never be bluebloods." Unlike those built in the junkyard-scrounging days of yore, she explained, these newer street rods "have no past, no history, no pedigree."[91] Others, especially restoration enthusiasts who happened across street rodders in parking lots or at old-car meets, complained that the modern street rod was an inauthentic mongrel precisely because it incorporated up-to-date conveniences. Defending the street rodder's approach, Will O'Neil of *Street Scene* wrote in an early 1981 column that he had a friend "who rebuilds old cars as a hobby and restores old pre–Revolutionary War homes for a living. He puts late model engines and hydraulic brakes in his old cars and, you know what, he puts electricity, air conditioning and flush toilets in those old homes."[92] Livability, much more than authenticity, was the key to the street rod's appeal.

Yet authenticity was precisely what street rodding aimed to recover when it was in its infancy in the 1960s. Thus, during the 1980s, 1990s, and 2000s, several splinter niches emerged among those who felt that street rodding had lost its way in the fiberglass- and comfort-obsessed 1970s—among those who felt, more precisely, that the street-rod movement had not gone far enough in its attempt to resurrect the spirit of 1930s, 1940s, and early 1950s hot rodding. Some therefore built "repro rods," nicely finished Ford roadsters with flathead V-8 engines, manual transmissions, and period-correct speed equipment.[93] Others, far more radical in their quest for authenticity, built "rat rods." Like repro rodders, rat rodders rejected the inclusion of automatic transmissions, air-conditioning, and anything else that wasn't period correct, but they also categorically rejected the notion that the pursuit of perfection in fit and finish was compatible with a desire to return to hot rodding as it once was. This was because the original rodders back in the "good old days" drove cars that were often very rough to look at, and they did so precisely because they valued the pursuit of performance and speed above all else, including aesthetics. Accordingly, early rat rodders spent all of their time and money on the way their cars ran and did very little to spruce up the way they looked. Over time, serious rat rodders actually came to value cars with rust, primer, and other signs of wear and tear, typically known as patina, so much so that many began to deliberately cultivate a junky look when finishing a car. Somewhat ironically, that is, in seeking to radically devalue their hot rods' aesthetics they actually came to value a particular counteraesthetic, the rusty and beat-up look.[94]

That they did not wish to drive cars that looked like junk did not mean that more mainstream street rodders were any less enthusiastic about the use of parts from salvaged cars than hot rodders, customizers, restoration enthusiasts, import and sports-car fans, and even rat rodders. One didn't need to advertise the lowly origins of what one drove, in other words, to be an avid and competent junkyard scrounger. For though they differed in their aims, they all were skilled at snatching cars and parts from the oblivion of the crusher and the shredder. And to a man, regardless of what their finished projects looked like, none of them were ever keen to see the things they rescued heading back from whence they came—hence their opposition to public and private efforts to eliminate so-called clunkers. Hence as well the advent of the specialty insurance sector.

"Insurance for People Who Love Cars"

Conventional automobile insurance policies, long required of motorists in most U.S. states, rely on an array of actuarial tools, depreciation schedules, market-value tables, and claims-adjuster judgments in order to set their annual premiums and to determine how much money goes to whom when a claim is filed. For the most part, these complex formulations boil down to a handful of simple truths. First, the more one drives, the more one pays in premiums. Second, the worse one's driving record, the more one pays. Third, the higher the rate of accidents and thefts in one's neighborhood, the more one pays. Fourth, the more one's car is worth, the more one pays. Fifth, and most important for our purposes, as a car ages and accumulates mileage, it depreciates in value. This means that a four-year-old Toyota, for example, will generally be worth less than a three-year-old Toyota. As a result, insurance firms will tend to pay out less on claims involving four-year-old cars than they will on those involving three-year-olds of the same make and model. Thus, if a three-year-old car with a "depreciated market value"—also known among insurers as an "actual cash value"—of $6,000 were involved in a collision requiring $4,500 in repairs, an insurance firm would tend, all things being equal, to pay for those repairs. If a four-year-old car of the same make and model but with a depreciated market value of $4,000 were involved in the same collision, however, chances are that the insurance company would deem the car a total loss. Rather than cutting a $4,500 check to a body shop, it would settle the claim by making a $4,000 payment to the car's owner. The firm would then recoup some of this money by selling the wrecked vehicle to an auction company like Copart. Depending on the extent of the car's damage and what we might call its "real-world market value," it might then be purchased by a salvage yard, or it might be bought instead by a used-car dealership willing to repair it for resale.[95]

The key to all of this lies in the difference between a given car's "actual cash value," as assigned by an insurance company, and its "real-world market value"—the actual amount of money it would take to buy a car in the same condition, precollision, on the open market. In our hypothetical scenario, the four-year-old car that the insurance company valued at $4,000 might in fact have been worth $5,000 on a used-car lot or in the pages of *Auto Trader*. Consequently, many late-model cars written off in accidents or other mishaps have actually ended up back in use. Consider what happened in the wake of Hurricane Katrina. In late 2005, 2006, and 2007, a number of cars that were damaged when New Orleans flooded were repaired and cycled back into use, rather than being scrapped for parts, after they were formally written off. Although totaled cars with salvage titles often end up back on the streets, a number of "Katrina cars" were laundered through the title systems of more than one state, which enabled them to be sold with clean rather than salvage titles. The resulting outrage from used-car buyers who felt that they'd been duped into buying flooded cars—which often suffer from intermittent electrical problems, mold, and other issues—led to calls to transform the National Motor Vehicle Title Information System, created in the 1990s to address automobile theft, into a mandatory clearinghouse for tracking titles across state lines.[96] But that's another story, for another time. What matters here is what made those "Katrina cars" possible in the first place: the difference between a car's "actual cash value" and its "real-world market value," a difference that in this case made it profitable for shops and used-car lots to purchase totaled cars for repair and resale.

Since at least the 1970s, this difference has also mattered a great deal to many of the enthusiasts featured in this chapter. In 1997, Bob Chuvarsky of *VW Trends* explained why: "Most of us have known someone with a vintage Volkswagen who dropped, say, a couple thousand dollars in a motor, another few thousand for body and paint, not to mention slapping down more cold cash for the interior, wheels, suspension and so forth. Then, after investing all that time, money and sweat came the disappointment of finding out that after the VW was wrecked the insurance company assigned the car an actual cash value considerably less than what had been invested in it."[97] The situation was even worse for those with cars that did not neatly mesh with their insurers' expectations regarding vehicles of a certain age, make, and model. How does one assign a value to a 1936 Ford street rod, for example, that was based on a reproduction chassis and a fiberglass body from the 1980s but had an engine from the 1950s, brakes and suspension parts from the 1960s, and interior appointments from the 1970s? Or to a dune buggy loosely based on a 1960 Beetle but that shared only a handful of its parts with other Beetles that had not been modified in a similar manner? More broadly, was there any way to insure a valuable but seldom-driven antique, classic, or special-interest car without breaking the bank?

Among the first with an answer, at least for those with fully restored, period-correct antiques, was the James A. Grundy Agency of Fort Washington, Pennsylvania. Beginning in 1949, this firm offered inexpensive policies to cover thefts and collisions involving antique cars, provided they were only used for "hobby purposes and exhibition" rather than for everyday transportation. This made it possible for the growing ranks of restoration hobbyists to affordably insure their cars so that they could legally drive them to shows and meets.[98] The J. C. Taylor Antique Auto Insurance Agency of Upper Darby, Pennsylvania, addressed the thornier matter of the difference between "actual cash value" and "real-world market value" when it joined the fray in 1966 by offering "agreed value coverage" to its customers. This innovation, which others were quick to copy, enabled enthusiasts to factor in the actual replacement costs of their valued rides when drawing up a policy.[99] Soon other firms entered the market, including Dempsey and Siders, American Collectors Insurance, and Condon and Skelly in the 1970s and Hagerty Classic Insurance in the 1980s.[100]

As the competition heated up, new policies emerged with more options for the owners of modified cars, as well as more flexibility in terms of the types of situations covered. Early policies often required the owner of a covered car to remain in its presence at all times, for example, unless it was securely locked away in a garage. This made it impossible to stop at a diner while on a Sunday drive or to wander around at a car show out of sight of one's insured vehicle. Condon and Skelly and Hagerty both addressed this by dropping their "attendance clauses" in the 1980s and 1990s.[101] Other policy innovations that emerged as the specialty insurance sector grew included provisions allowing covered cars to be driven as much as the owner wished, without mileage limits, as long as they were never used for the daily commute or other business purposes.[102] By the end of the 1990s, it was possible for enthusiasts to buy policies from specialty insurance companies that were inexpensive and flexible and allowed them to enjoy their vehicles without fearing the otherwise inevitable financial pinch associated with a collision or a theft.

Nothing could prevent a cherished car from being utterly destroyed in an accident, of course, but that was never the point. Instead, the object was to recover one's full investment should the unthinkable occur. Among those at the upper end of the collector-car market—the rarified realm of multimillion-dollar Ferraris and Duesenbergs—one's financial investment was well worth protecting for its own sake, especially since many who owned cars worth that much money bought them first and foremost as investments rather than expressions of pure enthusiasm.[103] On the other hand, among gearheads with more down-to-earth cars, recovering one's investment in an insurance settlement was as often as not a means to an end, a way to get behind the wheel of a similar antique, classic, or special-interest car as soon as possible and with minimal financial loss.

As for the mortal remains of their totaled vintage cars, sometimes these were bought back by the policyholders themselves, to be rebuilt with their settlement money. Sometimes they were snapped up at insurance auctions, often by other hobbyists who were seeking a challenging project.[104] Others wound up in salvage yards instead, but even this did not mean certain doom. For among the many damaged antiques, classics, and special-interest cars that have been sold into the salvage industry over the years, a select few have been fortunate enough to end up in a type of wrecking business less obsessed than most with things like turnover rates and the price of scrap: an enthusiast-specialty junkyard.

3

Arizona Gold

Enthusiast-Specialty Salvage Yards, 1920s–2000s

Even among die-hard old-car restoration enthusiasts, *Hemmings Motor News* has never exactly been a page-turner. Unlike *Cars and Parts, Classic and Sports Car, Car and Driver,* and other popular monthlies, it rarely features more than a handful of articles, photographic spreads, and how-to guides per issue. Instead, from the 1980s to the present, the typical issue of *Hemmings* has served as a clearinghouse for all things automotive—the "world's largest antique, vintage, and special-interest auto marketplace," in the words of its publishers.[1] As such, it consists almost entirely of classified listings and quarter-, half-, and full-page advertisements, all sorted by car model and year. From humble origins as a thin circular in 1954, *Hemmings* grew in tandem with the collector-car hobby during the 1960s and 1970s. By the early 1980s, with its five-hundred-plus tracing-paper-thin pages of business and classified listings, *Hemmings* looked and felt a lot more like a copy of the local *Yellow Pages* than anything else. And like the *Yellow Pages*, few would ever bother to read it from cover to cover.

But if they were to do so, they might notice a peculiar set of statistical anomalies associated with its salvage-parts and salvage-business advertisements. Each month scores of automotive wrecking yards from across the country take out spots in *Hemmings* to connect with subscribers who might need anything from engine blocks and differential gears to salvaged body panels and interior trim for their projects. Likewise, individuals place thousands of classified listings in this printed marketplace each month, several hundred of which invariably deal with salvaged odds and ends—an unrestored heap here, boxes of parts there. The growth of a robust reproduction-parts business in the 1970s, 1980s, and 1990s has meant that many parts no longer need to be scrounged from wrecks in this way, but even today there are parts on most old cars that simply cannot be bought new; they are, in the parlance of the gearhead, "unobtainium" or "NLA" (no longer available). This is why so many wrecking yards and individual enthusiasts list their salvaged goods in *Hemmings* and why so many others eagerly flip each month to the sections covering their favored marques and eras.

The statistical anomalies in all of this relate to the geographic distribution of those who advertise. Consider the issue of *Hemmings* published in December of 2003. That month, 538 enthusiasts took out classified advertisements listing salvaged components, parts cars, and the like. Of these individual listings, 59.1 percent came from just eight states: Arizona, California, Colorado, Florida, Idaho, New York, Pennsylvania, and Texas. And of these eight, only Pennsylvania and New York accounted for percentages roughly comparable with their share of the total fleet of vehicles registered in the United States. In other words, among the top advertisers, only these two states' ratios of in-use cars to available out-of-service cars and parts came close to what one would expect, 1:1.[2] Texas tracked slightly low, while the others tracked high, often by rather large margins. Whereas Idaho was home to just 0.4 percent of all cars registered in the United States in 2003, for example, its enthusiasts accounted for 3 percent of individual *Hemmings* salvage-parts classifieds that December. California, then home to 13.8 percent of registered vehicles, accounted for a whopping 21 percent of these classifieds; Florida (6.3 percent of registered cars) for 11.9; Colorado (0.7 percent) for 6.3; and Arizona (1.5 percent) for 3.5. Notice the overall pattern: with the exception of Florida, all of the leading states that tracked high were located in relatively arid regions, and with the exception of Texas, all of the leading states that tracked low or at par were not.

Precisely the same distribution held among salvage-business listings. Ninety-two wrecking yards from twenty-five states paid for a total of 393 advertisements and classified listings in that December issue of *Hemmings*.[3] Of the states represented, nine tracked low vis-à-vis their share of the nation's cars (Alabama, Georgia, Indiana, Massachusetts, Ohio, Maryland, Minnesota, New Jersey, and Virginia), five at close to 1:1 (Florida, Illinois, New York, North Carolina, and Washington), and eleven tracked high (Arizona, California, Colorado, Idaho, Kansas, Oklahoma, Oregon, Pennsylvania, South Carolina, South Dakota, and Texas). With the exception of

Washington, all of those that tracked low or at par were Rust Belt, south-eastern, or northeastern states, and with the exceptions of Pennsylvania and South Carolina, all of those that tracked high were more arid western, southwestern, or northwestern states. Moreover, 261 of the issue's total of 393 salvage business advertisements came from arid places, 159 from Arizona and California alone. In short, if you were to sit down and read this issue of *Hemmings* from cover to cover, you would likely come away with the impression that individual and for-profit automotive salvage tends to take place in those parts of the country that are either arid or salt-free in the winter—places, that is, where cars and parts are less susceptible to rust.[4] But you would only be half right.

Like gas stations, parts stores, and repair shops, automotive salvage businesses are distributed fairly evenly across the United States by vehicle population. Although their methods have evolved a great deal since they first emerged en masse in the 1910s, their basic role in the maintenance of American automobility has not: they absorb the many irreparably damaged cars and light trucks that we discard each year, salvaging components from them for reuse in other vehicles before dispatching the rest to the scrap-metal industry. As a rule, used parts are almost always cheaper than new replacements, and they also tend to fit better. Consequently, professional mechanics, body shops, and do-it-yourselfers from Bangor to San Diego and from Key West to Seattle have long relied on wrecking yards to keep our motors running. And from Maine to California, differences in climate notwithstanding, one is rarely more than a few miles from a salvage yard that can supply a serviceable wiper motor for a late-model Dodge, a transmission for a late-model Ford, or an alternator for a late-model Toyota.

Specialty salvage businesses that cater specifically to automobile enthusiasts are a different matter. These are the sorts of firms that advertise in *Hemmings*, and as the aforementioned data from 2003 suggest, they are far less evenly distributed than ordinary salvage yards. Since the 1950s, when specialty wrecking yards first emerged in significant numbers, states in the West, Northwest, and Southwest have been overrepresented in the business compared with those in the Rust Belt, the Deep South, and along the East Coast. New York, for example, home to an average of 6.6 percent of the nation's cars between the 1950s and the early 2000s, accounted for only 3.4 percent of specialty yards in that time frame. Twenty-one others also tracked low, nineteen of which are Rust Belt, Deep South, or East Coast states. On the other hand, eighteen tracked high, including fourteen relatively arid states like California and Montana (11.2 and 0.36 percent of the national fleet of cars but 14.2 and 1.14 percent of the country's specialty salvage yards, respectively). But Arizona takes the crown. Its vehicle population has averaged only 1.1 percent of the national total over the last sixty years, but its arid environment has been home to more than 6.3 percent of the specialty salvage business.[5] Not for nothing have so many salvage-yard advertisements in gearhead periodicals bragged of "Arizona gold," "rust free, desert dry Arizona" inventories, "Arizona absolutely rust-free" parts, "Arizona rust-free sheetmetal," or "absolutely A-1 rust free 1950s–80s auto and truck parts from Arizona."[6] By contrast, not a single firm has ever tried to sell "Chicago gold" or "rust-free Cuyahoga parts." Instead, firms that operate in places like Illinois and Ohio generally do not mention the sources of their parts at all, and when they do, it is invariably because they too have "Arizona gold"—salvage vehicles imported from the arid West.[7]

But why are "rust-free, desert dry" parts of such significance? More precisely, why is the rust-free mantra so much more important to the specialty salvage-yard business, in comparison with the general automotive salvage industry, that it skews the geographic distribution of the sector? Are there other ways in which the composition of the enthusiast-specialty wrecking business differs from the mainstream salvage industry? How does the day-to-day operation of a specialty business compare with that of a more general-purpose salvage yard? Most broadly, what exactly is an "enthusiast-oriented specialty salvage yard," and why does it matter? This chapter addresses these and other questions by exploring the development of the specialty salvage business from its origins in the 1920s and 1930s, through its maturation as a sector in the 1950s, 1960s, and 1970s, and on to the beginning of its denouement in the 1990s and 2000s. Specialty wrecking is a sector of the economy that most have never heard of, but its operations are nevertheless important to the broader story of automotive salvage in the United States and to the broader history of technology in use (and, in this case, technology in reuse).[8] For the story of the specialty salvage business adds not simply another layer of nuance to our understanding of the circle of automotive life, from the factory to the showroom to the driveway to the junkyard to the scrap yard to the scrap mill and back to the new-car factory.[9] Instead it demonstrates, in conjunction with the stories of reuse and bricolage detailed in the preceding chapter, that under certain circumstances that circle can be broken. For more than fifty years, in fact, much to the chagrin of the scrap-metal dealer and the delight of gearheads of all stripes, the only things that cycled out of a number of salvage yards were rust-free parts, destined for new lives not by way of the shredder, furnace, and new-car factory but through the capable hands of car enthusiasts.

The Rise of the Enthusiast-Specialty Salvage Yard, Part 1: Hot Rods and Sports Cars

Wrecking yards with particular specialties have existed since the salvage industry was young. By the early to mid-1920s there were yards that focused on common brands, like Ford, or on less common ones, like Franklin.[10] Shortly thereafter, salvage yards catering specifically to the needs of automobile enthusiasts began to emerge as well. Among the earliest, and quite possibly *the* earliest, was a small wrecking yard in Bell, California, opened by a man named George Wight in 1923.[11] At first his business was not unlike any of a number of other yards scattered in industrial suburbs coast to coast. But in the middle of the 1920s, his yard began to attract a very specific type of customer: the hot rodder.

Southern California hot rodders took to salvage yards as important sources of high-performance parts and ideas in the mid- to late 1920s, and as they began to frequent Wight's yard, he decided to take a strategic turn. For Wight was wiser than most when it came to these early hot rodders. He knew all about their dry-lakes races and their modified roadsters, and he also knew that brand-new aftermarket equipment could be costly. As a result, he began to carefully screen the wrecks that rotated through his inventory, pulling a high-performance carburetor from an Auburn here and an aftermarket cylinder head from a modified Ford there. Soon he began to actively seek out wrecked performance cars and modified vehicles as well, and by the time the stock market crashed in 1929, his small salvage yard in Bell was well known among shoestring-budget racers as a place that understood their needs.

As the Southern California hot-rod scene matured in the 1930s, so did Wight's business. By 1931 he had begun to sell new speed equipment from his wrecking-yard office along with select used parts. A few years later he began to produce his own equipment under the brand name Cragar, and by the end of the 1930s the new-parts business was vastly more significant to him than the used-parts trade had ever been. Although George Wight passed away in 1945, his business would live on in the hands of one of the local hot rodders who frequented his salvage yard in the 1930s, Roy Richter. With Richter at the helm, Bell Auto became one of the most successful high-performance firms of the postwar era, producing Cragar speed equipment as well as helmets and other safety gear under the Bell name. What matters the most for our present purposes, however, is that low-budget racers never forgot what they learned about used-parts hunting in the 1920s and 1930s at places like George Wight's yard. More to the point, some of these enthusiasts even followed Wight's lead into the specialized business of used hot-rod parts themselves. Lee's Speed Shop in Oakland, California, for example, a well-known postwar enterprise, had its roots in a 1930s secondhand hot-rod-parts business that Lee Chapel operated in Los Angeles after getting his start several years earlier as a yard hand at another operation.[12] Although many of them ultimately gravitated toward the new-parts trade, men associated with hot rodding like George Wight and Lee Chapel were the first to blaze the specialty used-parts trail.

Far more important to the emergence of the specialty salvage business, especially as a more widespread trade, was the advent of the import and sports-car hobby in the 1950s and 1960s. Those who drove imported cars during these years often found it difficult to locate parts. This was particularly true for those who drove makes that were rare in the United States, like DKW and Simca, but it was also true for many of those who owned more common ones like Porsche, Jaguar, MG, and Triumph. As a result, used parts from wrecked examples quickly came to be a vital resource among those who preferred these smaller and sportier cars to the much larger cruisers then available from the American industry. And as the demand for secondhand parts for these imported cars grew, several coastal salvage yards began to pay a lot more attention to them.

One such yard was Sherman Way Auto Wreckers of North Hollywood, California. Its inventory during the early 1950s consisted almost entirely of late-model domestic cars, but by the middle of the decade its operators had begun to cater to growing local demand for parts for European imports. By

1956, the year it placed its first advertisements in the enthusiast-oriented *Sports Cars Illustrated* and *Road and Track*, it had sufficient rare-bird inventory to offer parts for "Jaguar, Porsche, VW, MGTD, MGTC, MGTF, Alvis, Anglia, Austin, Healey, [Co]unsul, Hillman, Jowett, Morris, Renault, Rover, Singer, Sunbeam, [and] TR-2" makes and models. Its owners did not treat their foreign and domestic inventories any differently, at least at first, because space was scarce on its one-acre lot. This meant that, foreign and domestic alike, only a handful of cars were kept in the yard at any given time, while the rest were taken apart and their useful components stowed away for future sale. At the end of 1958, however, the operation acquired a larger lot in the same neighborhood, changed its name to Grand Prix Auto Parts, began to stock several hundred cars at a time, and quickly settled into its new role as a foreign and sports-car specialist.[13]

A similar shift occurred at Frank and Al's on Long Island in Westbury, New York. This yard also got its start in the business as a typical domestic-make yard, but it gradually assumed a foreign-specialist strategy during the 1950s. Its operator's rule of thumb was to send any car, foreign or domestic, to the scrap yard if two customers in a row looked it over and found nothing of use. Nevertheless, Frank and Al's attracted phone- and mail-order business from points far afield, and drop-in customers often managed, when "swarm[ing] over [a] picked-clean shell," to "come up with something that everyone else missed."[14] Others on the East and West Coasts soon joined the foreign and sports-car specialty business as well, including Jack's Auto Parts of suburban Washington, D.C., which specialized in British makes beginning in the late 1950s; All Auto Parts of North Hollywood, California, which operated "2 large yards devoted exclusively to imported cars" in the early 1960s; and Road and Track Auto Parts, a foreign and sports-car haven in the early to mid-1960s in Sun Valley, California.[15] Import and sports-car owners living in the heartland, often far from the nearest dealers that handled their makes and models, soon had access to local specialists as well when businesses like the Foreign Car Parts Mart of Cleveland opened their gates in the early 1960s.[16] Similar yards proliferated coast to coast during the 1970s, 1980s, and 1990s, including yards devoted to air-cooled Volkswagens, water-cooled Volkswagens, Porsches, BMWs, Fiats, Jaguars, Mazdas, and Nissans/Datsuns, as well as British cars, French cars, Swedish cars, and European cars in general.[17] From humble East and West Coast origins, the foreign-specialist salvage trade was here to stay.

Here a word of typological clarification is in order. Few of the foreign-specialist yards mentioned above dealt only with cars belonging to their areas of expertise. And not every foreign-car salvage operation necessarily catered to automobile enthusiasts—those mentioned above certainly did, for reasons discussed below, but not every yard with a similar inventory operated in the same way. Likewise, neither those Ford-specialist yards of the 1910s and 1920s nor George Wight's hot-rod yard of the 1920s and 1930s, nor the many muscle-car, old-car, and other specialty yards covered later in this chapter dealt only with their respective foci. Finally, very few general-purpose automotive salvage yards have failed over the years to handle at least the occasional classic, exotic, or other niche-market vehicle. With all of this in mind, what exactly *is* an "enthusiast-specialty salvage yard"? For the purposes of this analysis, an enthusiast-specialty salvage yard, whatever its particular area of expertise, fits one or more of the following criteria: coverage in secondary sources and enthusiast periodicals listed it among the yards that openly catered to one or another group of automotive hobbyists; the yard self-identified as enthusiast-oriented in its advertisements; or the yard was sufficiently attuned to (and capable of serving) at least part of the gearhead market that it chose to advertise in enthusiast-oriented periodicals. By these measures George Wight's yard was likely the first, while the foreign specialists of the 1950s marked the beginning of an era of more widespread specialty activity. But in terms of sheer numbers, energy, and cultural and geographical significance, nothing matched the advent of the old-car salvage yard.

The Rise of the Enthusiast-Specialty Salvage Yard, Part 2: Antiques, Classics, and Special-Interest Cars

Through the early 1950s, when automotive restoration was still a relatively new pastime and had yet to become a broad-based hobby, there were no old-car specialty salvage yards per se. Instead, those owners of antiques, classics, and special-interest automobiles who needed parts could simply go to a general-purpose wrecking yard and, according to *Motor Trend's* Robert Gottlieb, "purchase radiator cap ornaments, motometers, and other miscellaneous paraphernalia for old cars for junk prices." Bloated salvage-yard inventories, especially in the 1930s and very early 1940s, meant that parts were plentiful no matter what the make or vintage of the car in question. Complete project cars were also abundant. As late as the early 1950s, according to Gottlieb, "there were many fine classics to be purchased in unrestored condition for as little as $250," and "those in excellent shape rarely sold for more than $500." Even allowing for a bit of nostalgic exaggeration—Gottlieb's piece dates to the early 1960s, when the prices for unrestored and restored cars alike were on the rise—he was not far off the mark when he described the period through the early 1950s as the "good old days" for restoration hobbyists.[18]

As early as the middle of the 1950s, however, old-car parts had begun to grow noticeably scarcer. This was due in part to the organized scrap drives held across the country during World War II, which eliminated many older wrecks from salvage yards as well as barns and backyards. It was also due in part to Korean War–era directives that mandated higher rates of inventory turnover among salvage yards, further eroding the stock of prewar cars, as well as the related factor of the still-inflated price of scrap steel in the late 1940s and early 1950s.[19] It was also due in part to marketplace realities, for as the 1950s wore on, the cars of the 1910s–1930s were no longer quite as profitable for general-purpose salvage yards to keep on hand as were their later-model wrecks from the late 1940s and early 1950s. Finally, the growing scarcity of old-car parts and project cars during the mid-1950s was also due in part to the growth of the restoration hobby itself—to simple competition within the avocation's growing ranks.[20] All of this was especially true of antiques and classics. Special-interest cars, either for parts, restoration, or hot rodding, could still be found in salvage yards across the country in the mid-1950s, although they were gradually becoming harder and harder to find as well.[21] In this environment of growing scarcity, some began to set up operations specializing in old-car salvage.

Among the first to do so was a certain Mr. Schulte of Youngstown, Ohio. A collector of classic cars, he opened Schulte's Auto Wrecking in 1952 specifically to cater to those, like himself, who were trying to keep the high-end cars of the 1920s and 1930s on the road.[22] Michael J. McManus of Gardena, California, also got into the business at an early date by way of his personal hobby. A collector of antique and classic cars and parts, McManus noticed in the early 1950s that their value was rising and that they were harder and harder to come by. Soon he quit his day job, bought a local salvage yard, and revamped its three and a half acres into an antique, classic, and special-interest operation. His gamble paid off, for in the mid- to late 1950s the antiques and classics that he favored continued to grow ever more scarce. By the early 1960s, "his business [had] grown to a point where he [was] nationally known," as Robert Gottlieb noted in his aforementioned article on the state of the old-car hobby, and every day "inquiries [arrived] from all over the United States and many foreign countries seeking information as to the whereabouts of a given part."[23]

Albert B. Garganigo, the owner of an antique-car museum in Princeton, Massachusetts, dealt with similar inquiries in the early 1950s. Although his main endeavor was the restoration of vehicles for display in his museum, Garganigo also maintained a "boneyard" of parts cars and other wrecks both for his own use and in case "a fellow enthusiast wants to swap parts or a car." Garganigo's boneyard wasn't exactly a business, that is, but instead a private junkyard for his own use and for the occasional barter with like-minded collectors.[24] Others also hoarded private caches less for profit than for kicks. Perhaps the most well-known, at least among enthusiasts, was B. J. Pollard of Detroit. He began collecting unrestored antiques and other cars on the grounds of his contracting business in 1939, amassing 250 by 1946 and close to 700 by 1974. Although Pollard's collection was private, he did allow car clubs and the press to tour the site on more than one occasion.[25]

Others followed suit, especially after World War II. Willie Milligan began accumulating old and otherwise unwanted cars on a flat piece of land outside Eugene, Oregon, in 1951, for example, and although he never parted any of them out,[26] he did occasionally sell them whole to other hobbyists. But it was his own enthusiasm, rather than the potential for profit, that actually motivated him to assemble his private outdoor "museum." "Most of the older cars he has have been there for many years," *Cars and Parts'* Mark J. Hash reported in a story about Milligan and his hoard in 1991, "as it takes a little coaxing to convince Willie to part with some of his old friends."[27] Charlie Cleveland of Golden, Mississippi, was much more willing than Milligan to sell complete cars from his private collection, but he was also motivated less by profit than enthusiasm. "In the 1950s, Cleveland and his father started picking up cars, bringing them home, and just parking them," Randoll Reagan explained in a *Cars and Parts* feature on Cleveland in 2000. "The result," Reagan continued, "is that he now has a rather large collection of older cars, some of which are fairly scarce."[28] Mel and Ken Jackson of Yakima, Washington, also amassed a private collection, mostly of Hudsons, Nashes, and Cadillacs, beginning in the 1950s. Over the years, Mel worked as a mechanic, a body man, and an insurance adjuster, all of which gave him inside information that led to many of his purchases. Although they did sell complete cars from time to time, the Jacksons' collection, like those of Milligan and Cleveland, was primarily a private cache.[29]

But the bulk of the old-car specialty business did consist of firms, like those of Schulte and McManus, that were deliberately established to profit from the sale of used parts. Unlike those of Schulte and McManus, however, most of these enterprises did not begin as old-car specialists per se. Instead, they grew into that role during the 1950s and 1960s as the special-interest hobby expanded. Dwaine Bridley of Glenville, Minnesota, launched his wrecking business in 1954 and, after shifting to a larger lot in 1958, simply let his inventory of older models grow. "I like the old cars and never crushed them," he later explained, "although I do recycle newer vehicles and have my own crusher."[30] Pearson's Auto Dismantling and Used Cars in California's Central Valley had similar origins. After opening as a general-purpose wrecking yard in 1952, the business moved twice during the 1950s before settling in Mariposa, where until the end of the 1980s its owners never crushed a single car. Consequently, vehicles from the 1940s, 1950s, and 1960s piled up at Pearson's, earning the yard a solid reputation among the growing ranks of the special-interest restoration hobby.[31] The same was true of Stewart Criswell and Sons of Nottingham, Pennsylvania, which opened in 1945 and gradually developed into an eighteen-acre old-car haven during the 1950s and 1960s. "We never crush the old stuff," Alan Criswell told Joe Sharretts of Cars and Parts in 2001, because "you never know what someone will need."[32] A similar approach turned Elwood's Auto Exchange, founded in 1951 in Smithsburg, Maryland, into a massive old-car specialist over the next couple of decades. Elwood's operators not only refused to crush their own old-car inventory, but they also actively sought out and purchased old-car inventories from other, more mainstream salvage yards that would have been less reluctant to crush and scrap them.[33]

Dozens of others grew into the old-car specialty business during the 1950s,[34] and scores more followed in the 1960s and 1970s. By the 1980s, coverage of individual old-car yards began to appear in American enthusiast periodicals like Old Cars Weekly, Cars and Parts, and Hot Rod, as well as Canada's Old Autos. But Cars and Parts went further than most. Beginning in the early 1980s, it ran feature articles on enthusiast-oriented salvage yards in nearly every single issue that it published. By the mid-2000s, in fact, these stories had become sacrosanct, a form of coverage so popular among the magazine's readers that it was unthinkable not to include one each month. Brad Bowling, the title's managing editor, found this out the hard way: "A while back, I had to remove some pages of editorial at the last minute to make everything fit between the covers that month, and a salvage yard story was exactly the right size. When I mentioned the need to make room in that issue to *C&P* publisher Mark Kaufman, he told me to drop anything *but* the salvage yard feature! I did as advised, and made a mental note to never put out an issue without photos of rusty cars sitting in shoulder-high weeds with lots of parts missing."[35] Part of the reason for the features' popularity was their sheer utility in the ongoing quest for cars and parts among antique, classic, and special-interest enthusiasts. "I ended up buying a car from your April '98 issue," one reader wrote to the editor in the fall of 1998. "I bought the 1960 Dodge Dart Seneca, picture number three. Thanks!"[36] Walter and Mary Jane Lula of Doc's Auto Parts, a salvage yard featured in a *Cars and Parts* story, also wrote late in 1990 to thank the magazine for generating many calls and mail-in queries, which "gave us and the business a boost."[37]

As Bowling's remarks suggest, however, another reason for the popularity of *Cars and Parts*' salvage-yard features was that many readers actually enjoyed flipping through pages and pages of rusting heaps. "I want to say that the salvage yard feature is my favorite part of the magazine," one reader wrote to the editor in 1995, adding that "the monthly ritual in our family is for my wife to take the magazine out of the sealed wrapper and cover the photo captions with 'Post-Its.' I then try to identify the vehicles by year, make and model."[38] Others did the same, and over the years more than one wrote in to report that the magazine had misidentified a particular wreck in one of its features.[39] For other readers, the salvage-yard coverage in *Cars and Parts* served as a virtual supplement to their own junkyard-scrounging adventures. "I enjoy your salvage yard section," wrote one, who added that he had "always liked poking around old junkyards."[40] Others concurred, and their sentiments reflected but the tip of an iceberg: an emotional attachment to salvage yards, as well as a passion for wading through their weeds, mud, and rust, was widespread among gearheads of all stripes by the end of the twentieth century. For now, however, let us turn to the ways in which enthusiast-specialty businesses managed their affairs.

Day-to-Day Operations in the Specialty Salvage Business

Foreign- and sports-car specialists rarely operated any differently than run-of-the-mill, general-purpose salvage yards. But some of them did grow out of private collections in the manner of the old-car yards of Albert Garganigo, Willie Milligan, Charlie Cleveland, and Mel and Ken Jackson. For example, Welsh's Jaguar Enterprises of Steubenville, Ohio, today a well-known new- and used-parts depot, owes its origins to a private collection of XK-series roadsters from which its founder, William Welsh, began to market salvaged parts during the 1960s.[41] Others evolved into old-car yards during the 1960s, 1970s, and 1980s as Jaguars, Porsches, Volkswagens, MGs, Triumphs, and other erstwhile cutting-edge imports gradually became special-interest restoration candidates themselves.[42] But apart from purchasing strategies geared toward the acquisition of particular brands and types, most foreign- and sports-car specialists were indistinguishable in their day-to-day operations from ordinary wrecking yards. Some were even on the cutting edge of mainstream salvage trends. Foreign Auto Specialists (FASPEC) of Portland, Oregon, adopted a streamlined, dismantle-and-store approach when it opened its doors in 1966, years before such an approach would once again become common among ordinary salvage businesses. Other foreign specialists were early adopters of telex interlinks, computerized inventory management, e-mail queries, and Internet-based marketing as well.[43] On the other hand, substantial deviations from mainstream salvage practices within the enthusiast-specialty sector tended instead to occur among old-car firms. More specifically, old-car specialists often operated outside the bounds of normal wrecking practice in five key ways.

The first involved the economics of the business. Simply put, profits in the automotive salvage industry as a whole are typically tied directly to rates of inventory turnover, like most other volume retail sectors. Accordingly, the more wrecks a yard can bring in, part out, and crush for scrap in a given period of time, the more money it will make. However, for reasons hinted at above and discussed at greater length below, the nature of the old-car specialty business prevented most of those within it from pursuing rapid rates of inventory turnover. Instead, what they brought in tended to stay on their properties for extended, often indefinite intervals, reducing their profitability over the long term. This was especially the case, as *Cars and Parts'* Joe Sharretts lamented in 2007, when "people who know of classic salvage yards keep the locations to themselves because they don't want to have competition for parts." Secrecy such as this, Sharretts continued, was penny-wise but pound-foolish, because old-car salvage operators "need to sell parts in order to make a living. If an owner can't sell enough parts from the classics, crushing is sure to begin."[44]

Nevertheless, some were able to sustain their operations on the basis of old-car parts and systems over the long run. Ed Summar's exclusively 1950s and 1960s yard in Bradford, Arkansas, thrived for more than twenty years, while Joe Pierce's six-acre yard in Miamisburg, Ohio, remained in business for more than thirty and Leroy Walker's forty-acre old-car yard in Beulah, North Dakota, for more than forty.[45] As late as the 1980s and 1990s there were newcomers as well, entrepreneurs who were convinced that they could make an old-car-only venture work. After buying a thirty-year-old yard in Umatilla, Florida, in the 1980s, Tom Lee drew up plans to crush anything made since 1980 so that his new business, which already consisted almost entirely of vehicles from the 1950s, 1960s, and 1970s, would become exclusively old-car oriented.[46] Connie Toedtli of Vancouver, Washington, entered the trade in a slightly different fashion. Long the owner of a body shop, Toedtli "accumulated quite a number of old cars" over the years, and "by the mid-80s he had quite a stash." This prompted him to go through the difficult process of obtaining the necessary permits and licenses to use his collection of wrecks as the basis of an old-car yard. The result of Toedtli's efforts, All American Classics, opened on ten acres in 1989. By 1996 it was bringing in additional old cars at a rate of forty to sixty per month, and it had grown to cover eighteen acres.[47] Wayne Haynes of Plant City, Florida, enjoyed similar results after buying an established yard, crushing its modern inventory, and restocking it with 1930s–1970s foreign and domestic wrecks.[48] Even in the rust-prone Northeast, a part of the country where old-car yards were scarce to begin with and were growing ever more so with each passing year, there were some who bucked the trend. Ron Tluchak of Winslow, New Jersey, opened an old-car yard in 1996 that remained entirely devoted to 1920s–1960s vehicles through the decade of the 2000s. Tluchak's business strategy, neatly summarized by Joe Sharretts in a 2008 *Cars and Parts* feature, was simple: "nothing modern is ever brought in."[49]

But these were the exceptions, not the rule. Most old-car yards stayed afloat by supplementing their antique, classic, and special-interest salvage incomes with sales of parts from newer, more mainstream wrecks. In fact, for a number of old-car specialists, the newer-model parts trade was less a supplement than a mainstay, a stable stream of income that allowed them to remain active in the much less profitable vintage-parts sector. Just under half of R. K. Cochran's Speedway Auto Wrecking in Dacono, Colorado, a 1950s and 1960s specialist, were newer, mainstream models that helped pay the bills. Likewise, Morrisville Used Auto in Vermont, well known in the Northeast by the early to mid-2000s for its stock of 1920s–1960s wrecks, reserved more than one-third of its acreage for newer models; Engler's Used Autos and Parts, known among special-interest enthusiasts as "one of Pennsylvania's best kept automotive secrets" for its extensive stock of 1940s and 1950s vehicles, ran a newer-model operation on 20 percent of its land; Albin Avenue Auto Salvage in Lindenhurst, New York, a vintage muscle-car specialist, ran a second, later-model yard a few blocks from its main location; and Moriches Used Auto Parts of Center Moriches, New York, a fifteen-acre operation, maintained a rolling inventory of approximately 2,000 newer wrecks to supplement the income from its 750-odd vintage collection. The same was true of Hidden Valley Auto Parts of Maricopa, Arizona; Martin Supply of Windsor, Colorado; Gilly's Auto Wreckers of Placerville, California; and countless others.[50] This of course raises an obvious question: why bother? Why did so many devote even a fraction of their businesses to the old-car parts trade when newer, mainstream models were not only more abundant but also far more in demand—and thus far more lucrative when broken up for parts?

The answer is simple: vintage-car enthusiasm on the part of the yards' owners. This enthusiasm also drove the second and third key ways in which old-car specialists differed from mainstream salvage yards. The second, and most straightforward, involved the types of inventory they maintained. Whereas those who ran ordinary salvage yards deliberately sought out common and profitable later-model wrecks, those who ran old-car yards more often sought out those that tugged at their heartstrings. This was particularly true of several yards already noted in this chapter—those of Schulte, McManus, and Elwood, among others—as well as John Casiello's Unique Auto and Truck of Jackson, New Jersey. A vintage Cadillac specialist, Unique Auto opened in 1998 on thirteen acres formerly occupied by a late-model yard. After crushing out the extant inventory, Casiello "then went on a buying spree, following any and all leads about any and all old cars he might hear about." The result, by 2006, was a yard filled with vintage cars, close to one-quarter of which were his "favorite marque, Cadillac." Similarly, Leroy Martinez and Eric Christensen of L&M Used Auto Parts in Alamosa, Colorado, confessed in a 1995 feature on their business that they both "ha[d] 'soft spots' for the old timers" they amassed on their forty-acre property. From time to time, specialty-yard owners also managed to pass their "soft spots" down to a second generation. "My dad had a love for cars that stayed with him for his entire life," Pat Perillo Jr. told Dennis David of *Cars and Parts* in 1999. Consequently, his father's primary goal as the owner and operator of Mt. Tobe Auto Parts in Plymouth, Connecticut, was "to keep older cars on the road" by supplying parts from his wrecks. Until the day he died he never crushed a single one of his vintage wrecks, his son went on to explain, adding that since he took over the business in 1990, he and his personnel "ha[d] continued this tradition, and all of our cars have been here since the beginning."[51]

The third key operational difference between old-car specialists like Mt. Tobe and more mainstream salvage businesses involved the ways in which they were managed. By the end of the 1970s and especially during the 1980s, late-model yards were rapidly embracing a more streamlined approach to the business. Sometimes this entailed the adoption of computerized inventory management. Sometimes it involved bringing system and order to the filth and chaos of an open yard. On occasion it even meant the end of open salvage altogether in favor of dismantling and used-parts warehousing. Many old-car yards participated in some or all of these shifts, especially the rise of digital inventory-management systems in the 1980s, e-mail- and web-based marketing in the 1990s, and environmentally friendly practices in the 2000s.[52] But more than a few clung instead to more traditional approaches. For some, this meant keeping mental track of where each older wreck was located—and what parts it still had left to yield—rather than relying on computer databases. At Joe Pierce's yard in western Ohio, for example, Joe alone knew what he had on hand and where; the same was true of Pat Perillo's business in Connecticut.[53] On the

other hand, R&S Auto Parts in Moffett, Oklahoma, did have a database of
its newer-model inventory, but when it came to its stock of several hundred vintage vehicles, its owner, Dwayne Roberts, relied instead on his
own memory.[54]

Other old-car yards retained a more traditional, open-yard feel simply
by letting chaos reign within their gates. Uneven and unordered groups
of wrecks were scattered willy-nilly on Perillo's lot, for instance, and at
Fredericksburg Auto Salvage in Virginia, Southwest Auto Wrecking in
Las Vegas, and Stewart's Used Auto Parts near Hartford, Connecticut.[55]
Likewise, weeds, trees, and underbrush were allowed to obscure entire
cars at Ed Lucke Auto Parts in Glenville, Pennsylvania; Leo Winakor and
Sons in Salem, Connecticut; and Graveyard Auto Parts in Coldwater, Ohio.[56]
Not coincidentally, these overgrown, chaotic old-car yards fit well with an
outlook, long common among old-car enthusiasts and other gearheads by
the 1980s, that actually celebrated mud, weeds, and rust as vital to the
thrill and romance of the "salvage yard treasure hunt." "Although many of
the cars are easy to spot," Dennis David wrote in a 1997 feature on Winakor's, "there are some that are so overgrown with brush that it's hard to
tell if a car is there. While this may irritate the modern parts hunter," he
continued, "old car enthusiasts will take heart in this 'hunt for buried treasure.'"[57] Other contributors to popular periodicals also waxed enthusiastic
about the joys of hunting for parts in overgrown yards, and calendars,
posters, and large-format books featuring photographs of vehicles rusting
away in old-car yards sold well.[58] Among gearheads increasingly inclined
to cherish the thrill of the old-car-parts hunt, those salvage yards that
bucked the neat and tidy, streamlined-salvage trends of the 1980s, 1990s,
and 2000s were apt to garner favor. For this reason, many old-car yards
not only allowed their customers to pull their own parts but also let them
browse at will, with no particular car or part in mind. Little wonder, then,
that Schulte's Auto Wrecking featured a large "playground" sign above its
gates.[59]

The fourth key way in which old-car specialists differed from more
mainstream yards also involved inventory management. More specifically,
it hinged on what Alan Criswell meant when he explained that older cars
were never crushed at his family's yard because "you never know what
someone will need."[60] Whereas late-model yards have always had the luxury of calibrating their inventories so that they stocked common parts for

Figure 3. The upper B-pillar area of the front door frame of a 1967 21-window Volkswagen Microbus that sat outside for more than two decades. Over time the rubber seal for the skylight window above the door frame grew hard and cracked. Although it has since been replaced, as shown in the image, the damage was already done: for many years, this worn-out seal allowed water to enter the structure of the door frame, rusting it from the inside out. New replacement parts have long been available for virtually every body panel for these vehicles, but the rust on this particular example was highly unusual and thus required a donor section, or "clip," from a junkyard Microbus that was otherwise in much worse shape. (Author photograph.)

use in common types of repairs on common cars, old-car yards were often premised on the fact that automotive restoration can be an unpredictable endeavor. This is especially true of body panels, which often rust in particular areas—wheel arches, rocker panels, floor pans—but can also rust in other, less predictable ways. A vehicle parked outside on a hill each day might eventually develop rust in unusual places based on how rainwater happened to pool upon its surfaces, while another with bad sunroof or window gaskets might rust through its frames and pillars from the inside out (fig. 3). Although the same kind of corrosion often occurred once a vehicle wound up in its final resting place in an old-car yard, even the most wretched wrecks still had the potential to yield precisely what a particular customer needed. "No matter how crunched and rusty they are," Betty Francy of Fredericksburg Auto Salvage explained to *Cars and Parts* in 2006, "there's something somebody can use from them."[61] Consequently, many old-car yards routinely cut sheet metal to their customers' specifications or sold them uncommon odds and ends. "We sell a lot of cutouts," or clips, Jeff Hoctor of Hidden Valley Auto Parts in bone-dry Arizona explained in 1990.[62] The same was true at Turner Auto Wrecking in Fresno, California, whose owner advised *Super Chevy*'s readers in 1997 that the best way to go about ordering a clip was to fax or mail a drawing indicating precisely where the cuts should be made, including a bit of extra metal in every direction to allow for easier welding.[63] However, offering cut-to-order service like this required old-car yards to keep many of their old wrecks on hand for extended periods, and not every yard was willing or able to do this.[64] For a lot of others, though, a crusher-free approach was the only business model that made sense.

The fifth and final key to the old-car salvage sector is that with which this chapter began: geography. Old-car salvage yards existed primarily as repositories for restoration parts that were difficult if not impossible to acquire elsewhere. These often included rare or otherwise oddball sheet-metal clips, air cleaners and other intake parts, exterior chrome trim, emblems, lenses, glass, and interior parts—especially radios, ashtrays, lock buttons, door pulls, and the like. The elements were never particularly kind to most of these things no matter where a car resided. Lenses would eventually haze up and crack, as would safety glass. Depending on where the car spent the bulk of its time, its interior appointments would either dry out and crack or grow moldy, and its electric devices would either corrode internally or their plastic buttons would whiten and crumble. As a result, old-car yards were all alike when it came to many items, regardless of where they were located.

The exceptions to this rule of variegated automotive decay were few, but significant: sheet metal, ferrous chassis parts, chrome, and emblems. Both in areas with harsh and humid winters requiring liberal applications of road salt and in areas with high humidity in the summer, chrome would easily pit, emblems would corrode, and chassis parts and sheet metal would develop deep and porous rust. Vehicles in the Northeast and the aptly named Rust Belt were particularly susceptible due to winter salt. "We're a bit weak on sheet metal," Bob Adler of Adler's Antique Autos in Stephentown, New York, admitted in the late 1990s. Joe Pierce of A Lotta Auto Parts in western Ohio agreed, adding that as a result of his area's harsh winters, most of the otherwise unbent cars in his yard were only good for parts and could never have been put back into service even by the most enterprising of enthusiasts.[65] Following the vehicles of the Northeast and the Rust Belt were those of Hawaii, where high humidity and salty ocean breezes were the chief culprits.[66] Vehicles in the Southeast tended to fare a lot better, although high humidity and frequent rainfall often took their toll there, too.[67]

Those in the Northwest, the Mountain West, and the western plains were much less susceptible to rust and chrome rot, because much of those regions is salt-free in the winter (urban cores excepted) and much more arid than the Northeast, the Rust Belt, tropical Pacific islands, and the Southeast. "The climate here isn't particularly wet in the same sense one associates with other areas of the Northwest," wrote *Cars and Parts'* Doc Howell in a feature on Classic Auto Parts in Coeur d'Alene, Idaho, in 1993, adding that "fortunately, the state doesn't salt the highways," either. Michael G. Beda was more succinct in a 2007 feature on Speedway Auto Wrecking of Dacono, Colorado: "the dry Eastern Colorado air keeps the Demon Rust at Bay." On the other hand, precisely because the climate in these areas tended to preserve rather than destroy, cars that would have been considered restoration candidates back east were often torn apart out west. As Todd Toedtli of All American Classics in Vancouver, Washington, explained in 1996, "People on the East Coast would cry with some of the cars we cut up."[68]

But by far the best protected were those that spent the bulk of their time, both in life and in the automotive afterlife, in the warm, salt-free, and arid Southwest. "We finally know what dry is, or at least where it's found," explained Ken New of *Cars and Parts* in a 1990 feature on Hidden Valley Auto Parts. He then went on to detail how doors, hoods, and other sheet-metal sections pulled from the cars in this Maricopa, Arizona, salvage yard were "as rust free as the day they were stamped."[69] Numerous articles in enthusiast periodicals from the 1980s, 1990s, and 2000s concurred, advising readers back east to buy their project cars out west or, at the very least, to make use of the treasure trove of rust-free parts scattered across the region's salvage yards.[70] Their readers took heed, buying everything from sheet-metal clips to complete cars from western yards and shipping them back east, up north, and even abroad.[71] For those who were looking to turn back the clock, the lure of the West was strong indeed.

Consequently, over the last sixty-odd years, old-car yards in particular have tended to concentrate disproportionately in the western plains, the Northwest, and especially the Southwest. In fact, if we adjust the data from this chapter's introduction to focus only on old-car yards, we find that nearly every arid state that was overrepresented in the enthusiast-specialty business as a whole was overrepresented to an even greater extent when it came to old-car yards. This was true of Arizona, Colorado, Idaho, Montana, Nebraska, Nevada, Oklahoma, Oregon, South Dakota, Washington, Wyoming, and to a lesser extent North Dakota, Texas, and Utah. Alaska, Kansas, and New Mexico had slightly smaller shares of the old-car trade than they did of the specialty business as a whole, while California, for reasons discussed below, stood alone among arid states in having a *substantially* smaller share of the old-car business than it did of the enthusiast salvage sector more broadly. (To be clear, in absolute numbers California did have far more old-car yards than any other state, but its share of the old-car business was substantially lower than its share of the enthusiast salvage sector as a whole.) Meanwhile, many eastern, Rust Belt, and southeastern states also had larger shares of the old-car sector than they did of other kinds of specialty salvage businesses, but among them only Delaware, Iowa, Maine, and Wisconsin were actually overrepresented in the old-car business in an absolute sense—vis-à-vis their shares of the national fleet of cars.[72] But when it came to old-car salvage yards, by far the most significant statistical deviations involved Arizona and California.

Arizona, with an average of just 1.1 percent of all vehicles registered in the United States over the last sixty years, has nevertheless been home to better than 7 percent of old-car salvage specialists. Among states overrepresented in the old-car business, this is more than double the combined share of its nearest two rivals, Wisconsin and Washington, both of which averaged far more registrations than Arizona over the last six decades. Even among states overrepresented in the business, that is, Arizona stood alone—hence the number of firms located in the Grand Canyon State that either featured "Arizona" in their names or listed it prominently in their advertisements, as well as the number of old-car firms back east that imported rust-free cars and parts from arid Arizona.[73]

California, on the other hand, has been home to far more enthusiast-oriented salvage yards over the last sixty years than any other state. It has also been home to far more old-car yards than any other, including Arizona. Statistically, however, California was overrepresented only within the enthusiast-specialty business as a whole, but not within the old-car sector in particular: against 11.2 percent of the nation's cars over the last six decades, the Golden State has been home to 14.2 percent of specialty wrecking yards but only 11.2 percent of old-car salvage businesses. In other words, California has long been overrepresented in servicing the used-parts wants and needs of the enthusiast community as a whole, but not among old-car hobbyists. How could this be, especially since its arid climate is nearly ideal for the preservation of iron and steel?

The answer is twofold. First, since the beginning of the 1990s, California authorities and business interests have crushed and shredded far more older cars than any other state in "clunker" or "scrappage" operations. Designed to remove older-model cars from the road for good, most of these California programs, explored in greater depth in the final chapter of this book, did not allow parts to be recovered from the older cars they handled. In this way, the regulatory environment has actually discouraged old-car salvage in the Golden State for the last twenty-five years. Second, and more important, real estate near California's major population centers has long been exorbitantly priced, so much so that salvage yards there tend to run a bit small. As a result, its wrecking businesses often have less room for extensive inventories than those in other states, limiting the ability of the old-car specialists among them to follow the low-turnover, never-crush approach favored by most antique, classic, and special-interest yards.[74] Nevertheless, 11.2 percent of the nation's old-car yards is still a significantly larger share of the pie than any other state's, even Arizona's. "California cars" and "California rust-free sheetmetal" thus were every bit as familiar to the enthusiasts of the late twentieth and early twenty-first centuries as those that hailed from Arizona.[75]

"Salvage Yard Alert"

During the 1990s and 2000s, old-car yards across the country began to close. One by one they either auctioned off their inventories or called on scrap dealers to dispose of what they had in mobile shredders. Enthusiasts shuddered as yards well known to generations of high-performance and restoration enthusiasts called it quits: Hap Gemmill's in 1990, American Auto Salvage in 1991, Clark Auto Parts in 1993, the Swartz Salvage Yard in 1994, Classic Auto Recyclers in 1998, Leo Winakor and Sons in 2000, Harmon Auto Wrecking in 2001, Lakeside Recyclers in 2005, Wheels of Time in 2008, and Hauf Auto Supply in 2009, among others.[76] Some of these yards closed because of what their owners saw as an increasingly unfriendly regulatory environment. "The largest hurdle to operating a successful salvage yard in today's business climate," Gene Hauf explained to Ron Kowalke of *Old Cars Weekly* as he shuttered his old-car yard in Oklahoma after sixty-three years,

> is the increasing level of regulation being handed down from both the government and the insurance companies that underwrite policies covering yards. It means slogging through a mountain of administrative paperwork to comply with DMV, EPA, and IRS rules. Due to the ever-tightening insurance regulations, it's gotten to the point that customers are no longer allowed on yard property. This essentially defeats the purpose of having collector-era vehicles and their parts for sale as old-car parts are definitely a see-and-touch-before-purchase commodity.[77]

Sixteen years earlier, Clark Auto Parts of York, Nebraska, auctioned off its inventory for similar reasons, whereas many others closed instead because of eyesore, zoning, and other disputes with local officials.[78]

Far more often than not, however, the old-car yards that closed in the 1980s, 1990s, and 2000s did so not as a result of federal and state environmental regulations, nor because of local meddling, but for simpler reasons. Sometimes scrap-steel prices were the culprit. During the 2000s, skyrocketing scrap prices tempted the owners of many old-car yards with less than satisfactory foot and phone traffic into crushing out their inventories and closing up shop. This was especially true along the East Coast, but nowhere was the trend more pronounced than in northwest Alabama and Mississippi. "In my visit to yard after yard," *Old Cars Weekly*'s Ron Kowalke wrote following a salvage-yard excursion across the Deep South in the late 2000s, "it was pretty much the same story. The owners admitted that there wasn't much demand for old-car parts in Mississippi, so crushing what vintage iron that remained in their salvage yards, due to its weight, just made good business sense."[79] Exurban sprawl and the attendant rise in real-estate values on the suburban fringe were to blame in several other cases. So too were the loss of property leases, sudden illnesses, or on occasion a simple desire to retire from the business after many years of pulling greasy parts from rusted heaps.[80] Other yards "went modern," as it was often put, bowing to the economic pressures of the business by crushing out their old-car inventories in favor of much more profitable later-model wrecks. This happened even at yards run by enthusiasts. At Smith Brothers Used Auto Parts in Westminster, Maryland, older-model cars had been the focus for more than five decades when Sean Donohue bought the place in 2001. "Although Donohue loves classic cars," Joe Sharretts explained in a *Cars and Parts* feature six years later, "economic reality [has] forced him to strip classics that had few parts left and send the shells to the crusher" so that he could bring in more of what his customers needed: cars built since 1980.[81] On a broader scale, competition from the reproduction-parts business gradually shortened the list of items once classed as "unobtainium." By the 2000s, for instance, it had been possible for many years to build a complete reproduction 1932 Ford street rod from scratch, using only brand-new parts, and it was fast becoming possible to do nearly the same for 1950s Chevys and several other cherished vintage models.[82]

But by far the greatest threat to the old-car salvage yard has been time itself. In part this is because most cars that are left in open yards will eventually return to earth no matter how dry the air or enthusiastic the yards' owners. More important, 1934 Chevrolet Coupes, 1951 Hudson Hornets, and 1967 Volkswagen Microbuses have long been out of production. Once all of the salvageable parts are picked from those few that still remain in old-car yards, there won't be many—quite possibly, *any*—to take their place. And when as a result the last of the old-car yards eventually closes its gates for good, it likely will not garner much attention from those who do not flip through *Hemmings* or *Cars and Parts* each month. But it surely will elicit a collective sigh from those who grew up at a time when old-car yards and other specialist enterprises were still vibrant elements of the automotive hobby. Indeed, clear signs of their resigned frustration at the growing scarcity of old cars, old-car parts, and old-car salvage yards have long been evident. Let us turn, therefore, to the language and practice of the old-parts treasure hunt, the mythology of the barn find, and the melancholy beauty that so many gearheads found in wrecking yards.

4

Junkyard Jamboree

Hunting for Treasure in the Automotive Past, 1950–2010

Through detailed features and brief pictorial spreads, automotive periodicals have covered out-of-service cars and salvage yards on a regular basis since the dawn of mass automobility. In the first decade of the twentieth century, *Motor, Horseless Age, Motor Age,* and *The Automobile* often speculated on the fate of out-of-date or "ancient" cars, some of which were scrapped but most of which remained in use in one form or another. During the 1910s, these same magazines began to feature salvage yards, describing how the Auto Wrecking Company of Kansas City or the Progress Auto Parts Company of Cleveland managed their affairs. They printed countless advertisements for individual salvage businesses, and they remained interested in the fate of older models, too, running brief spots on this or that "old-timer" still in use. An increasing number of how-to pieces also appeared. These were intended to help owners keep their cars on the road, or to help dealers refurbish and sell their stocks of used cars. During the 1920s, even as it gradually became a dealer-centered trade journal, *Motor Age* continued to run features on old cars still on the road. So did *Automotive Industries*, a manufacturer-oriented title descended from *Horseless Age* and *The Automobile,* as well as the trade-centered *Ford Dealer and Service Field,* a latter-day descendant of the *Fordowner.* But the big story during the course of the 1920s was the rise of the dealership-cooperative salvage yard, the new-car dealer's answer to the "used-car problem" in a number of American cities. During the 1930s, a trickle of human-interest stories centered on salvage yards, salvaged parts, and older-model cars continued to appear. By the time the United States entered World War II, however, the automotive press was geared entirely toward matters affecting new-car manufacturers and dealers.[1]

The resulting void would not be filled until the late 1940s and early 1950s, when a number of new publications geared specifically toward automobile owners were launched. These included general-interest titles like *Auto Age* and *Car Life*, as well as enthusiast-oriented magazines like *Speed Age*, *Hot Rod*, and *Motorsport*. As this happened, detailed coverage of salvage yards, salvaged parts, and older cars began to return. Some magazines ran stories intended to educate ordinary do-it-yourselfers on the ins and outs of buying parts from wrecking yards.[2] Others advised hot rodders on the things to look for when shopping for late-model V-8s for their projects, customizers on the art and practice of using salvaged parts to reshape and restyle their cars, and import and sports-car enthusiasts on the use of salvaged parts to keep their unusual rides alive.[3] Articles appeared as well on finished projects that began as hopeless wrecks or incorporated parts and systems pulled from cars in salvage yards. From the 1950s through the early 2000s, features like these remained common across the growing spectrum of niche publications.[4] So too did another angle on wrecking yards and salvaged parts that first appeared in the 1950s: coverage portraying the pursuit of certain parts and systems as a real-life treasure hunt.

Rare in the 1950s and 1960s, this treasure-hunting metaphor became more widespread during the 1970s and was endemic among gearhead publications and their readers by the 1980s and 1990s. Most often it was used to describe the search for valuable parts in open salvage yards. But from time to time it served instead to liven up discussions of antique, classic, and special-interest swap meets, as well as to bring color to narratives about the hunt for new old stock (NOS) components, original replacement parts for older cars that went unused for many years. At the same time, stories also began to circulate about rare or otherwise desirable cars discovered in salvage yards, in suburban backyards and garages, and especially in rural fields or barns. By the 1980s and 1990s, the resulting mythologies of the "barn find," the "junkyard rescue," and "the one that got away" were common, and enthusiasts of all stripes took to peeking into every nook and cranny just in case. Also widespread by the 1980s and 1990s was an important corollary of the treasure hunt and the barn find. For as they searched for hidden gems and forgotten cars, many gearheads developed an enthusiasm for

Figure 4. Automotive treasure rarely sparkles as found. This middle seat
for a mid-1960s seven-passenger Volkswagen Microbus—an "unobtainium"
component long difficult to locate *anywhere*, in *any* condition—was found
beneath a coffin in a mid-1970s van in a salvage yard near Atlanta, Georgia,
in the 1990s. (Author photograph.)

junkyards *as such* and for wrecked or otherwise deteriorating cars *as such*.
This was especially true among those who restored older models, but it
was also the case among high-performance enthusiasts, import and sports-
car fans, and others. The result, particularly in the 1980s, 1990s, and 2000s,
was a veritable flood of magazine features, books, posters, videos, and calendars celebrating not only the beauty of rusting relics and salvage yards
but also the connection to a bygone era that many felt when wandering
through fields of superannuated wrecks.

This chapter explores these interrelated developments. It begins with
an analysis of the origin and evolution of the treasure-hunting trope, followed by an examination of the culture and practice of the old-car swap
meet. It then delves into the lore of the junkyard rescue and the barn find
before ending with a detailed look at how and why a number of gearheads

of every stripe developed an affection for the traditional, open style of salvage operation over the course of the last forty-odd years. Along the way
this chapter demonstrates, above all else, that the timing associated with
all of this was no coincidence. From the 1970s through the 2000s, as
junkyard treasure hunts, barn-find quests, and the celebration of the
salvage-yard aesthetic peaked, open wrecking yards were closing coast to
coast, caches of NOS parts were becoming even fewer and farther between, and older cars themselves were growing more and more difficult to
find—especially time-capsule specimens. For the harder it became to actually locate hidden gems, the harder everyone looked, and the more spec-
tacular their stories of success began to seem (fig. 4).

"Salvage Yard Treasure Hunt"

The high-performance enthusiasts of the 1920s and 1930s were the
first to see more than just greasy parts and rusty scraps of metal when
they visited a salvage yard. Consider the early hot rodders who frequented
George Wight's wrecking business in Southern California. When they
dropped by and poked around his property in Bell, they saw neither hopeless junk nor merely serviceable parts. Instead they saw speed equipment
on damaged Fords and Chevrolets. They saw carburetors, wheels, and other
odds and ends on more expensive wrecks that could be adapted to their
own, more humble rides. Most important, they saw new ideas, new approaches, and new possibilities on totaled high-end cars, which shaped
their thinking as they continued the iterative process of modifying their
roadsters for better results on the dry lakes. Those who followed in their
footsteps in the early to mid-1950s saw the same things in their local salvage yards. Rodders of the 1950s eagerly waited for middle-class Americans to wreck their brand-new Chrysler 300s and Olds 88s so that they
could obtain their modern power plants at bargain prices for their prewar
roadsters, dragsters, and dry-lakes streamliners. Customizers of the same
period trawled through salvage yards as well, where they saw makes and
models of postwar cars yielding infinite possibilities in terms of custom
bodywork, as well as wrecked hulks waiting to be reborn as low-slung bou-
levard cruisers.[5] However, none of these enthusiasts ever described these
salvage yards and what they found within them in terms any different
from those of the scrap-metal specialist or mainstream used-parts seeker.
Early high-performance gearheads and car customizers pioneered the

practices that would eventually become part of the treasure-hunting trope among modification-oriented enthusiasts, but there is no evidence whatsoever that they ever thought of salvage yards in anything other than utilitarian terms, just like everyone else. They were simply sites where one could cheaply acquire parts, systems, and ideas—nothing more, nothing less.

Instead, the origins of the treasure-hunting metaphor lay with another group of gearheads, the antique restoration enthusiasts of the postwar period. Wrecking yards were often of limited use to these hobbyists by the early 1950s for reasons ranging from the scrap drives held during World War II to the limited production of many early models in the first place— as well as the ever-diminishing chances that after more than forty years, derelict antiques with salvageable parts might still remain in wrecking yards. Nevertheless, the search went on. Antique hobbyists scoured barns, garages, and salvage yards for useful parts and systems and for complete restoration projects, despite the fact that what they sought grew more elusive every year.

But it was this very scarcity, as well as the long odds that it generated, that led to the emergence of explicit "treasure hunting" imagery in the early 1950s. In an aptly titled 1953 piece—"Old Car Treasure Hunt Is On!"— the general-interest magazine *Cars* explained, beneath a lead photograph of a pile of mass-produced 1920s and 1930s cars in a salvage yard, that "America is in the midst of a new gold rush—a treasure hunt for antique cars." Although antique enthusiasts had no real interest in the cars piled up in that lead image, the article explained that an antique gem or two might still be found in mounds of cars just like them. The author, a Horseless Carriage Club of America (HCCA) official named Willard C. Poole, then went on to describe the ins and outs of hunting for antique cars and parts in salvage yards, rural sheds and barns, and other promising sites.[6] Some time then passed before this metaphor came into broader use. A contributor to *Auto Age* used it in a 1954 piece, describing a particular antique-car enthusiast's parts-gathering excursions as "treasure hunting trips,"[7] but after that it rarely surfaced again in print for more than two decades. Indeed, even as classic and special-interest hobbyists began to struggle somewhat in their ongoing pursuit of salvageable parts and cars in the 1960s, they rarely spoke directly about searching for salvageable treasure. Not until the 1980s would restoration hobbyists begin to mention "treasure" on a more widespread basis.[8]

In short, in much the same way that early high-performance and custom enthusiasts did not attach substantial emotional significance to the junkyard hunt, in spite of its importance to their hobbies, the antique-, classic-, and special-interest enthusiasts of the 1950s and 1960s did not normally celebrate the importance—indeed, the *centrality*—of the salvage yard to their pastime. They did not romanticize the parts-hunting experience, and they certainly did not waste their time gazing at hopelessly rusted relics. Instead, for them the junkyard was a utilitarian place. Moreover, especially in the 1950s, old-car enthusiasts almost always couched the question of a car's worth in terms of monetary values rather than sentimentality or restoration promise. Thus, when a reader asked *Motorsport* in 1955 whether it would be wise to rescue two derelict 1900s vehicles from a barn in Illinois—a question to which any enthusiast two decades later would have emphatically screamed "yes!"—the magazine responded with a matter-of-fact discussion centered on relative market values.[9] Even the more sentimental and treasure-oriented pieces of the 1950s were quick to note that completely worn-out and unrestored antiques and classics, with few exceptions, were virtually worthless except as sources of parts.[10] Twenty years later these sorts of opinions would have been incomprehensible to nearly everyone involved in old-car restorations. But for the level-headed enthusiast of the 1950s looking to get started in a still-developing hobby, they made good sense. Not until the early 1970s would old-car enthusiasts fully embrace the value, monetary and sentimental alike, of the rusted hulks and other salvaged parts upon which their hobby depended. But by then they had to compete for their hidden gems with a new class of junkyard treasure hunters: street rodders.

Street rodding began in the 1960s but truly came into its own in the 1970s. By then it had its own organization, the National Street Rod Association, founded in 1970, which coordinated local, regional, and national street-rod meets from coast to coast. It also received regular feature, editorial, and technical coverage in broadly circulated enthusiast periodicals like *Hot Rod* and *Popular Hot Rodding*, as well as in more specialized titles like *Street Rodder* and *Rod Action*.[11] And within these magazines, much more sentimental references to the humble origins of street-rod projects began to emerge. There were rags-to-riches stories—of a 1929 Ford rescued from an ignoble afterlife as a chicken coop; of a 1932 Nash literally dug out of the mud into which it had sunk over the years; of a 1932 Ford sedan delivery pulled from a wrecking yard as a birthday present for a family man who wanted to build a street rod; of an odyssey to the West Coast in search of a rust-free body for an East Coast project.[12] There were also articles urging enthusiasts to scour rural areas for abandoned prewar cars that could be transformed into street rods. For as Hot Rod's editors explained to their readers in the summer of 1971, "there's a lot more out there than meets the unsuspecting eye."[13] Even the authors of street-rod-oriented how-to pieces and technical articles began to describe their parts-gathering experiences as "junkyard jamborees," and to openly celebrate their many "junkyard finds."[14] It was at first a subtle trend. But as the hobby continued to grow during the course of the 1970s and 1980s, depictions of salvage-yard, backyard, and other discoveries as "jewels" and "prizes" and the quest for them as "hunts" became common.[15]

Street rodding was a nostalgic activity. This alone may account for the increasingly wistful sentiments of the street-rod press. But there were also very real shortages of certain cars and parts. The 1932 Ford roadster, for example, had always been the died-in-the-wool hot rodder's car of choice, and rebuildable examples had long been scarce by the time street rodding emerged.[16] Other favored cars like 1934 Ford coupes and genuine Model Ts were also growing scarce.[17] The same was true by the middle of the 1970s of V-8 engines worth rescuing from late-model wrecks. In the wake of the federal Clean Air Act of 1970, new-car performance fell precipitously in terms of power, reliability, and fuel economy as automakers settled for Rube Goldberg solutions to the challenge of meeting pollution-control guide-lines.[18] This meant that a street rodder working on a project in 1971 still would have had access to an abundance of low-mileage, high-output motors from the cars of the late 1960s. But the same would not have been as true for one working in 1976 or 1977. By then, desirable late-model engines—or at least, those that were ready to go right out of the wreck, without substan-tial modifications to unlock their emissions-strangled potential—were in-creasingly rare. Like finding a genuine 1932 roadster, locating a low-mileage, high-performance, bolt-in-ready engine from a late 1960s car increasingly seemed like finding buried treasure.[19]

Similar sentiments appeared during the 1970s among the devotees of a related niche, street machines. Typically defined as a 1949–1972 vehicle that has been modified for improved performance, street machines were the preferred choice among hard-core gearheads obsessed with horsepower, torque, and quarter-mile times. Because the street-machine scene included the very same muscle cars and pony cars that inspired the street rodding reaction, those who built street machines and those who built street rods often did not get along, especially during the 1960s and early 1970s.[20] But by the middle of the 1970s, they did share at least two important traits. First, both were obsessed with the past. For street rodders, it was a style of street performance and car construction they identified with the original hot rodders of the 1930s, 1940s, and 1950s, while for street-machine enthusiasts it was a style of street and strip performance they associated with the more recent cars of the 1950s and 1960s. Second, both relied on salvage yards. For with the end of the muscle-car era in the early 1970s, street-machine enthusiasts turned to salvage yards more and more frequently for suitable project cars and replacement engine and drive-train components. As this happened, the "junkyard jamboree" was quickly joined by "wreck-connaissance" missions, efforts to navigate "the salvage yard maze," and, of course, the venerable "salvage yard treasure hunt."[21]

Restoration enthusiasts soon joined in as well. Though members of this group were the first to speak in terms of "salvage-yard treasure" back in the 1950s, they were rarely guilty of romanticizing the sources of their projects or parts in the years that followed. Instead they were cautious, often willing to forget about the roughest or least valuable old cars they encountered. But by the early 1970s this had begun to change. "Oh, what glorious junk," exclaimed Menno Duerksen in 1970 while covering the massive swap meet held each year in Hershey, Pennsylvania, where enthusiasts from across North America gathered to buy, sell, and trade parts and potential projects. "Or 'junque,'" he continued, "as the aficionados are want [sic] to spell it now."[22]

Whether spelled with a "k" or the more pretentious "que," these early treasure hunters' use of "junk" flew in the face of contemporary trends within the salvage business. Like those who dealt in secondary materials in the nineteenth century, those who ran automobile salvage yards from the origins of the business in the 1900s through the early 1960s often had no qualms about the word "junk"—or, for that matter, "junkyard." But as federal and state authorities began to regulate the scrap and salvage trades on aesthetic and environmental grounds in the 1960s and 1970s, many within the business began to reject the use of "junk" and "junkyard" in an attempt to polish their public reputations.[23] Among enthusiasts like Duerksen, however, "junkyard" was and would remain simply another way of saying "salvage yard," "wrecking yard," or "automotive dismantler"—for them it was a synonym, not a slur. Indeed, gearheads often *preferred* to call their finds "junk," and their favorite hunting grounds "junkyards," even if they were aware of the industry's distaste for the terms. Perhaps this had something to do with the persistence of a rags-to-riches weltanschauung among certain enthusiasts, a worldview in which any of the more benign alternatives simply did not carry the emotional weight of "junk" and "junkyard."[24]

Whatever the reason, four years later Duerksen was at it again, describing "the little cry of triumph" heard intermittently among the crowd "when one of these eager searchers made his 'lucky find'" among the Hershey meet's "acres of exotic 'junk,' or 'junque.'"[25] Coverage of antique and classic-car swap meets and the restoration enthusiasts who attended them even made it into the pages of the performance-oriented *Rod and Custom*, where veteran street rodders John Thawley and Gray Baskerville bemoaned the effect on the price of rebuildables—and on the very definition of "rebuildable" itself—of the restoration crowd's emerging tendency to worship all things old:

> The price rise in relic rejuvenation has hit the sport like a Joe Frazier left. A rusted out, termite munched, shell of a pickup bed now commands many top dollars. What would have been considered a hopeless hulk of history, just a few years ago, is currently trailered in, trotted out and tagged with a price the size of our national debt. Beat, bullet riddled, broken and bent, original stock-type chassis, engine, body and accessory components are being peddled at such sky high figures that only the rich or foolhardy can afford to "pay" the game.[26]

If it was old, it was now worth saving, and the values attached to cars and parts that fit the bill were reaching new and, depending on one's point of view, potentially irrational heights.

More so than ever before, the old-car treasure hunt was on.

"Parts-A-Rama": Vintage Hoards and Seasonal Swap Meets

In 2002, *European Car*'s Kevin Clemens reflected on a sickness common among car enthusiasts: the irresistible urge to save everything, even worn or broken parts. "Sure, I don't own a Bugeye Sprite anymore," he explained, "but I really liked that car, so shouldn't I save a pair of slightly discolored front turn signal lenses in case I ever own another one?" *Car Craft*'s Matthew King concurred. "It's amazing what gearheads will keep just because they're too lazy, cheap, or stubborn to throw it away," he wrote in an editorial in 2000. "Maybe pathetic is a better word," he added, "for a guy who saves a Rochester 2-Jet carb for a truck he doesn't even own just because the bowl float might come in handy someday."[27] Clemens focused specifically on the "pack-rat mentality" of vintage sports-car fans, and King on the same phenomenon among American high-performance enthusiasts. But they could have written the same sorts of things about others. For as the hunt for automotive treasure intensified, a number of gearheads responded by amassing private hoards of parts, parts cars, and restoration candidates.

As we have seen, caches of this sort began to appear among antique hobbyists like B. J. Pollard of Detroit as early as the 1930s,[28] and over the next four decades, classic, special-interest, and muscle-car enthusiasts joined in. Pinky Randall of Houghton Lake, Michigan, for example, began collecting Chevrolet cars, parts, and memorabilia after World War II. Over the next forty years Randall accumulated more than a ton of used and NOS parts in his basement, garage, and outbuildings, as well as no fewer than thirty-five vehicles ranging from prewar coupes and roadsters to postwar gems, including a 1969 Corvair with just twenty-eight miles on the odometer. True to gearhead form, Randall also held onto "several carcasses from parts cars past."[29] Vernon Burks of Shelby, Ohio, preferred Model A Fords. After buying his first in 1958, he went on to amass more than forty more over the next four decades, some pristine, others less so, as well as an enormous collection of parts and memorabilia.[30] Like Pollard before them, Randall and Burks accumulated for their own enjoyment, not for public display. But all three did allow the automotive press to interview them, photograph their collections, and reveal their locations.

Others were more secretive, including a pair known only as the "Carolina Collectors." When Lee Beck of *Cars and Parts* visited their sizable private cache in 1990, all he was allowed to reveal in his subsequent story was that the cars were owned by a father-son restoration team, that they ranged from the 1910s through the early 1970s, and that they were located "somewhere in the warm climes of the Carolinas."[31] The same was true when *Hot Rod* ran a lengthy feature on a collection of nearly three hundred cars and several tons of parts in 2009. Although it did disclose that it was a man named Bob Regehr who had amassed the impressive hoard over the course of more than four decades, the magazine remained tight-lipped when it came to Regehr's zip code. Its closest hint? "Somewhere in the Midwest."[32] This was not the first time *Hot Rod* had extended such a courtesy. The previous year, it published a series of stories about private collections not unlike Regehr's in a special issue aptly titled "Hidden Treasure." In 2004, it also ran a picture of four hundred Mustangs in a field somewhere in Arkansas, sent in by the cars' owner, "Spike." A taunt from the secretive Spike accompanied the photograph: "Good luck finding 'em."[33]

From time to time, and much to the delight of treasure-seeking gearheads, private hoards like these came up for sale. When Don Schlag of Green Bay, Wisconsin, died in 2005, *Hot Rod*'s John Pearley Huffman reported that "he left behind 21 semitrailers stuffed with parts and cars" collected over nearly fifty years. Some things he bought locally, others on annual cars- and parts-hunting trips to the West Coast. Paranoid that someone might stumble upon his collection of hundreds of motors, more than a dozen rare cars, and tons of other parts and systems, Schlag not only packed the semitrailers himself but "parked them tail-to-tail on flattened tires to make opening them even tougher." Larry Fisette, another local collector, hit the "musclecar jackpot" when he bought the entire collection from Schlag's estate.[34] Likewise, over many years Henry Parrie of Great Falls, Montana, had amassed a private cache of close to three hundred cars and trucks, as well as salvaged parts and systems. "While he was alive," *Cars and Parts*' Dean Shipley reported, he welcomed visitors but "would not part with any of his machines, no matter how much money was offered." But everything went up for sale following his death in 1991, giving visitors the opportunity at last to leave the site without empty hands.[35] The same was true of Robert Glass of Angus, Minnesota. Upon his death in the mid-1990s, enthusiasts flocked to his properties, where the more than eleven hundred vehicles from the 1930s to 1970s that he had collected over the years were suddenly up for grabs.[36] Many other cherished hoards were dispersed to eager old-car hobbyists under similar circumstances during the 1980s, 1990s, and 2000s.[37]

Treasure-seeking gearheads also raided private caches of unused parts and accessories whenever and wherever the opportunity arose. When the owners of El Monte Ford, one of Southern California's oldest new-car dealerships, decided in 1982 to clear their warehouse full of NOS parts and systems, some of which dated back to the 1930s, they had no trouble finding a buyer. Valley Ford Parts, a vintage specialist in North Hollywood, snapped up the entire collection.[38] Other dealerships hired auctioneers to sell off their valuable NOS parts when their warehouses began to over-flow, as did the heirs of many private collectors who had amassed their own NOS hoards over the years.[39] As often as not, old but new-in-the-box parts like these gradually trickled into the inventories of specialty shops instead, which advertised them at a premium to old-car hobbyists obsessed with factory authenticity.[40]

Sometimes, the quest for NOS parts called for a bit more creativity than simply flipping through the classifieds. During the 1970s and 1980s, *Hot VWs'* Rich Kimball made several trips to Tijuana, Mexico, in search of unused parts for Beetles and other air-cooled models in old dealerships, repair shops, and private collections. Jeff Walters, another contributor to *Hot VWs*, did the same in the 1980s with journeys to Germany, Jamaica, Guatemala, and Venezuela.[41] Both found more than they expected. De-scribing a shelf full of NOS semaphores for 1950s Volkswagens that he and his companion found at an old dealership in Germany in 1984, Walters gushed that "there must have been 200 sets. We thought it was time for the 'big' one, had to take a nitro pill. Wow, there they were right in front of us."[42] Three years later, his heart intact, Walters found himself rummag-ing through another collection of NOS parts, this time in Guatemala. While bending down to pick up an ultrarare taillight component for a split-window Beetle, he "noticed two or three more laying half buried around it. I couldn't believe it," he later explained, "so I started digging 'em out. The more I dug, the more I found! By the time I had finished I had found a bag of 21."[43]

For his part, Kimball's first trip to Tijuana in 1970 involved an attempt to follow up on a juicy lead: rumors of a cache of NOS parts in a building occupied by a Volkswagen dealership in the mid- to late 1950s. Kimball and several others who joined him on the trip managed to locate not only what appeared to be the building in question but also the property's owner, a Dr. Bustamente. "I asked him if the building had once been a VW agency," Kimball reported, and Bustamente said yes. "By this time my knees were weak and I must have pumped a gallon of adrenaline through my system. I asked the doctor if there were any parts left. He said yes, but not many because they were only in business from 1954 to 1959. By this time I was about to go into cardiac arrest! I asked him if it would be possible to see them and he said he would come by in about a half an hour. It was the longest half hour of my life."[44] When Bustamente showed up and ushered him into the small storeroom, Kimball immediately spied rare taillight housings, distributors, and other NOS parts scattered on the shelves. But the owner was only interested in selling the entire collection at once, and in 1970 Kimball and his friends were unable to raise the necessary funds. Twelve years later Kimball returned, hoping to find the collection once again—and hoping as well that it was still intact, and still for sale. He did, and it was. This time he and a friend were able to meet the asking price, and one month later he returned to haul the treasure back to Southern California.[45]

The exotic adventures of Walters and Kimball were of course extraordinary. Far more common among old-parts seekers were trips to more mundane spots closer to home: salvage yards, used-parts retailers, and swap meets. Typically run by car clubs, often but not always in conjunction with antique, classic, or special-interest car shows, organized automotive swap meets began to appear in the mid-1950s. Among the earliest was the fall event in Hershey, Pennsylvania, launched by the Antique Automobile Club of America (AACA) in 1954 as an accompaniment to the group's annual car show. The Hershey meet was very small at first—seven vendors showed up to swap and sell their used and NOS parts in 1954—but it quickly developed into the premier event of its kind in the eastern United States. By the early 2000s, more than ten thousand vendors sprawled out over three hundred acres there each fall, and over time the "Hershey" name had become so well known that the organizers of other events tried to play off its fame. These included the "Little Hershey" event in Belvidere, Illinois, launched in 1966, and the Kyana Region AACA meet in Louisville, also known as "The Hershey of Kentucky," first held in 1969.[46] Other early events quickly became annual institutions as well, including the Mansfield Trading Bee, held in Massachusetts every spring since 1960; the Kalamazoo Mid-Winter Sell and Swap, a staple in western Michigan since 1964; the Portland Swap Meet in Oregon, held each April since 1965; the Antique Motor Car Swap Meet, an annual event at the Rose Bowl in Pasadena, California, first held in 1967; the Hoosier Auto Show and Swap Meet in Indianapolis, launched in 1967; the Southwest Swap Meet in Arlington, Texas, held each fall since 1968; Ontario's spring and fall Barrie Swap Meets, popular among Canadian enthusiasts as well as Americans living near the Great Lakes since 1971; and the annual events in Carlisle, Pennsylvania, first held in 1974.[47]

Most of the swap meets launched in the 1950s and the early to mid-1960s were put together by and for antique hobbyists. Then, beginning in the mid- to late 1960s, clubs for classic and prewar special-interest enthusiasts joined in. Sometimes they worked together. Antique and prewar special-interest clubs organized Arlington's Southwest Swap Meet, for example, as well as the Portland event in Oregon and the Hoosier meet in Indianapolis.[48] Others went it alone. Prewar special-interest clubs put together a number of events of their own beginning in the mid-1960s, as did hot-rod clubs, classic-car clubs, and import and sports-car clubs.[49] Postwar special-interest groups began to hold their own events in the 1970s as well, among the largest of which were the spring and fall Carlisle Swap Meets. Held not far from the site of the iconic Hershey show, the Carlisle events were launched in the fall of 1974 specifically to serve those who felt left out at other, prewar-only swap meets. By the 1990s Carlisle, like Hershey, had become a premier eastern-states event.[50]

During the 1970s and 1980s the number, size, and diversity of these meets continued to grow, and they came to play an important role in the life of the treasure-hunting enthusiast. But, like browsing a salvage yard, finding precisely what one needed in the rows and rows of "junque" and "classique"[51] parts often felt like searching for the proverbial needle in a haystack. This was especially true at larger shows like Hershey, which had six and a half miles of vendor rows by the early 1970s.[52] Some got lucky quickly, as Menno Duerksen explained in a piece about the 1974 Hershey meet, but "then there were those who trudged and trudged without success. As the hours and the days passed, the pace would slow, a slight limp might appear, [and] a look of weariness and despair would show." But even the most fatigued would push on, muttering to themselves as they rounded each corner that "maybe this will be my lucky row."[53]

Contributors to enthusiast-oriented magazines tried to help. Some advised hot rodders not to expect too many bargains at restoration-oriented swap meets. Others focused instead on the bargains that *did* exist. In a piece on the Labor Day Bug-O-Rama in 1983, Rich Kimball counseled his readers not to overlook the "junk boxes" many vendors set out, "from the bottom of [which] . . . blossom all kinds of treasures. Not only is this where the [vendor] put the stuff he didn't think was worth anything," Kimball continued, "but this is also where he put the stuff he didn't know"—the things he couldn't identify.[54] A few focused on the buyers and sellers themselves. Some offered guidance on negotiating a price from both points of view, while others looked specifically at who was drawn to swap meets and why. Many who sold or traded parts at these events were avid restoration enthusiasts looking to unload a few parts that were useless to them but potentially of value to others. Others were specialty shops and other serious used-parts retailers, while a few were simply gearheads looking to make a quick buck from parts scrounged from wrecking yards in anticipation of a major event. Among swap-meet buyers, most were hobbyists in search of odds and ends for specific projects, but some were there simply to collect rare and interesting parts for the fun of it. Others went instead to look for restoration projects or parts cars, both of which were readily available at local and regional swap meets.[55] Vendors offered more than 950 complete cars at the 2004 Springfield Spring event in central Ohio, for example, including everything from a half-restored 1936 Cord 810 to a rusty 1953 Nash Metropolitan. But perhaps the star of the meet was a 1933 Ford truck. Originally used as a utility vehicle by a Harley-Davidson dealership, the truck was "described as a 'Barn fresh, unrestored, all-steel, real patina'" example in tired but serviceable condition. Bob Stevens included a picture and description of the truck in a photographic spread on the Springfield meet for *Cars and Parts* later that year, and he even borrowed the seller's sales pitch for his title, "Barn-Fresh Finds."[56]

Neither the seller's choice of words nor Stevens's appropriation of them was a coincidence. For by the early twenty-first century, phrases such as "barn fresh," "unrestored," "all steel," and "real patina" were in widespread use among old-car enthusiasts. They signified a kind of holy grail, the so-called barn find: an unrestored but restorable vehicle pulled from long-term storage, rough perhaps but original in ways that apparently nicer cars, already refurbished at some point in the recent or distant past, simply were not.

"They're Still Out There, Somewhere"

Enthusiasts of all stripes have been pulling cars from salvage yards, urban lots, and rural fields for many years. Hot rodders often did so in the 1930s, 1940s, and 1950s, as did customizers in the 1950s and 1960s. In a more sustained way, restoration hobbyists have been doing so since at least the 1950s, street rodders since the 1960s, and import and sports-car enthusiasts since the 1970s, when their hobby became part of the special-interest fold.[57] The vehicles they found may have been preserved to some extent, especially if they came from arid western states, but few who rescued wrecked, rusted, incomplete, or otherwise time- and weather-worn cars from fields or salvage yards expected it to be easy to bring them back to life. Neither did most of those who found their projects sheltered in garages and other structures: far more often than not, these too were badly damaged or otherwise completely worn out. From time to time, however, some cars ended up in long-term, indoor storage long before they had a chance to be wrecked, and in some cases before any major repair or reconditioning work was ever done on them at all. After many years in storage, these original examples did of course deteriorate to a certain extent. Some developed rust, especially those stashed in humid spots. More than a few also suffered interior and electrical damage from rats and other vermin, while others came to be plagued by mechanical failures associated with long-term disuse. Nevertheless, because they were original and unmolested when they were squirreled away, stories of their reemergence years and even decades later came to be the stuff of gearhead lore.

Published tales of this sort were scarce throughout the first half of the twentieth century, and with good reason. Although the motoring press did begin to celebrate surviving "old-timers" and "ancient cars" on a limited basis in the 1910s and 1920s, it is worth remembering that automotive restoration as a hobby was almost unheard of in the 1910s and still rare through the 1920s. Even in the 1930s, the decade in which both the AACA and the HCCA were established, old-car restorations remained few and far between, and not until after World War II would the restoration of antiques, classics, and special-interest automobiles develop into a widespread pastime.[58]

As a result, tales of vintage-car discoveries only trickled into circulation prior to World War II. In 1912, *Motor Age* reported that Benjamin Briscoe of the United States Motor Company once found a homemade horseless carriage from the 1890s in a barn in San Francisco. Several years later, *Motor Age* also ran a story about a steam-powered omnibus built in 1855 by a pioneer of roadway locomotion, Richard Dudgeon, that was "gathering dust in a barn" in New York.[59] But perhaps the best example of a prewar barn find appeared in *Ford Dealer and Service Field* in October of 1930. That year, a car made by the Black Motor Company of Chicago was discovered by a new-car salesman during a house call. "This car is of ancient vintage," the magazine reported, but it "was stored in an old barn and forgotten for the past 20 years and never saw daylight in that time." After spotting it, securing its purchase, and hauling it back to the dealership, the salesman gave the car a quick mechanical checkup, after which it ran "like a clock." In the salesman's eyes, the best part was that this old car was something of a time capsule. "All of the original equipment is on the car with the exception of the top," he noted, adding that it "has been a wonderful advertising medium for us."[60]

After World War II, barn-find tales began to come into their own. In his pathbreaking article on the "old car treasure hunt," discussed at length above, Willard C. Poole of the HCCA summarized the situation as of 1953. His narrative centered on a hypothetical enthusiast, Joe. "If our friend Joe is like most," Poole began, "he will dream of entering some decrepit but very watertight barn and finding under a large dust cloth, a complete and handsome low-mileage car built in 1910 which only needs gasoline to start and new tires to roll." He went on to explain that finds like this were rare but not unheard of. As to how and why they might end up in storage in the first place, Poole offered a lengthy explanation. Some were simply parked when they broke down, and their owners never got around to fixing them. A few, especially early on, were "poorly designed in the first place, gave constant trouble," and were therefore set aside because they had little to no market value. Others were packed away rather than being traded in when their owners bought replacements, either because dealers would not give them enough in trade or because the owners decided to keep them for sentimental reasons. Finally, Poole maintained that some were "marooned" after being "taken onto islands, into logging camps, walled up in cellars" or barns, and so forth. Wrapping up on an optimistic note, he encouraged readers to get out there and join the hunt, because "there are still thousands of these vehicles yet to be uncovered."[61]

Five years later, *Sports Cars Illustrated*'s Ken Purdy was less sanguine. "As late as 1934," he claimed, "a Mercer runabout in good condition was sold for $75—and it was a tough sale, at that. But after World War II," he continued,

> the dam caved in with a rending crash, and the collectors began to collect in dead earnest. As more and more cars were bought up, and as the peasants holding most of them grew craftier and more grasping, the chase naturally got harder. . . . Time was when a week-end in the slum-like country would almost always turn up a car or two. You just drove around and poked your head in old garages and talked to the folks. Sooner or later somebody would remember that Jed Steegor had some kind of old car stashed away in his barn. You got a small boy to show you the way and set off, helplessly aware that long before your arrival, the bush-telegraph would have brought the word to Jed that a prime sucker was on the way, eager and loaded with folding [money]. . . . The really wise men began to think up dodges like post-card circulars: They sent post-cards to every postmaster, say, in the state of Pennsylvania, offering a small reward for leads on old cars. They advertised in obscure country week-lies. They staked out country drummers, men with years of experience in selling rural store-keepers, and plied them with truth serum. They tried everything but kidnapping, and I'm not really sure that somebody didn't have a go at that.

"If you want, say, a nice vintage sports car today," he concluded, "you can reconcile yourself to the fact that you're going to have a tough time finding it." For Purdy, the barn-find gig was already up.[62]

Fortunately for treasure-seeking gearheads, Purdy was wrong. "We are continually amazed at the desirable cars that are continually turning up," the editors of *Cars and Parts* reported in the summer of 1971, adding optimistically that "there are many barns left to look in and many a garage that has a desirable car in hiding."[63] Some intrepid old-car hunters found them one at a time. When Lauren Bowdish of Illinois bought two 1915 Finley-Robertson-Porter touring cars in 1929, he wrecked (and subsequently junked) one but drove the other in its original and unrestored condition until 1936. Then he simply parked the car in his garage, where it remained until an official from the Harrah's Automotive Collection in

Reno, Nevada, tracked it down in the mid-1970s.[64] Likewise, Rick Mader, a street rodder whose 1931 Ford pickup was featured in *Street Scene* in 1985, found his vintage Ford, originally converted into a street rod in the late 1960s, in a barn in the mid-1970s.[65] Others stumbled on small- to medium-sized collections as they poked around. Don Kear, an official with the U.S. embassy in Athens, Greece, happened upon a 1932 Ford, a 1930s Cord, and a 1953 Cadillac in an aircraft hangar on the outskirts of the city in 1983.[66] Closer to home, *Automotive Industries* reported in 1971 that "a treasure of antique automobiles—including one of the rarest vehicles known—has been discovered on a farm near Rossie, Iowa." The men who located and bought the cars reported that they "were stored in boarded-up barns and sheds and seven were found in a grove of trees." Among them was a rare 1905 Winton touring car, as well as a staggeringly original 1951 Chevrolet with only forty-three miles on the odometer.[67]

A lucky few found much more sizable collections. In the early 1980s, Herbert W. Hesselmann, a freelance photographer based in Germany, got a call from a friend, Guido Bartolomeo. Bartolomeo, who had recently moved into a cottage in rural France, reported that his new neighbor had a collection of some fifty-odd cars from the 1920s to the 1960s—including numerous Bugattis, Alfa Romeos, and Ferraris, as well as a handful of Lincolns, Chevrolets, Panhards, and others—that were stashed away in sheds and other outbuildings, as well as in the open. Hesselmann went on to photograph the collection for several enthusiast magazines, and in 2007 he collaborated with an automotive journalist, Halwart Schrader, on a lavishly illustrated book about the cars called *Sleeping Beauties*.[68]

In subsequent years, Bartolomeo and Hesselmann's multivehicle discovery proved to be truly exceptional: most of the barn finds made in the 1980s, 1990s, and 2000s involved one or two cars. But this did not make their stories any less appealing to gearheads, many of whom remained convinced that an amazing find with their name on it was still out there, somewhere. Reinforcing this belief were numerous barn-find reports in magazines from across the enthusiast spectrum and across the globe: a 1953 Kaiser Darrin in an open barn in Iowa; an all-original 1953 oval-window Volkswagen Beetle in a barn in Germany; an 11,235-mile 1972 Chevrolet Nova in a suburban shed in Delaware; a coach-built Rometsch sports car from the early 1950s in a building in Los Angeles; a 20,000-mile Chevrolet Chevelle in a storage unit in nearby Palmdale; a 1953 Ferrari 375 MM driven by Phil Hill in the 1954 Carrera Panamericana that was squirreled away in a storage container in the woods of Northern California for close to forty years; a 1970 Plymouth Superbird, once owned by Richard Petty, hidden for two decades in a garage in Lansing, Michigan; a 1937 Packard locked away in an old body shop in Leon, Mexico; a 1966 Aston Martin DB6 from a dry shed in the U.K.—the list goes on and on and on.[69] Websites documenting the phenomenon also helped to ratchet up the old-car-hunting craze, as did television shows like *Chasing Classic Cars* and, for motorcycle enthusiasts, *What's in the Barn?*[70] Popular books did too. Tom Cotter, perhaps the best-known barn-find chronicler in the United States, has published seven to date, including *The Cobra in the Barn* (2005), *The Hemi in the Barn* (2007), *50 Shades of Rust* (2014), and most recently, *Barn Find Road Trip* (2015).[71]

In addition to these titillating tales of well-preserved time warps discovered indoors, periodical coverage of treasure hunts involving rougher but no less desirable cars found moldering away in open fields, backyards, storage lots, and salvage yards also expanded dramatically in the 1980s, 1990s, and 2000s. In 1998, *Road and Track* published a remarkable story about a 1955 Ferrari 121 Le Mans Scaglietti Roadster that was discovered in a field in Southern California. The car, frequently raced in the mid- to late 1950s, was stashed away in a storage shed, where it remained for nearly thirty years. It was then relocated outdoors and placed beneath a semitrailer, between its front supports and rear wheels. "It was in this condition, in plain sight from a busy road," that a Ferrari enthusiast from the U.K. stumbled upon the rare gem in the late 1980s.[72] Similarly, in a story published in 1993, *European Car*'s David Featherston described how a Porsche aficionado, Jacques Gandolfo, tracked down a 1955 Speedster in Northern California. After hearing a story about an old Porsche in rough condition somewhere along Highway 101, "sitting on the back of a truck, covered with a faded, blue tarpaulin," Gandolfo set off. After driving less than half a mile, he found the car, exactly as described: "wrecked and undriveable," not to mention seriously rusty. But it was a genuine early Speedster, so he made a deal with its owner, took it home, and began a lengthy restoration.[73]

Somewhat less exotic but no less thrilling to the treasure-hunting gearhead was the tale of Tim Drachenberg's discovery of five desirable early Volkswagens in rural Gustine, Texas, in 1994. While on a long freeway drive across Texas, Drachenberg kept himself alert by making "side-to-side glances" along the way, "just in case he spotted that 'diamond in the rough.'" Sure enough, while passing through Gustine, "he thought he saw the back end of a VW bus, with a small rear window," parked next to a house. After getting off the freeway and circling back around, he knocked on the door and spoke with the bus's owner. The bus, a '62, was indeed for sale, as were four more vintage Volkswagens in a nearby sheep pasture. "Within ten minutes," Hot VWs' Robert K. Smith explained with palpable enthusiasm in an article about the cars in 1997, Drachenberg "struck a deal, and became the owner of all five VWs!"[74]

Similar tales leapt from the pages of high-performance and custom-oriented periodicals. In 1986, Hot Rod ran a lengthy feature on six custom cars from the 1950s that were owned by Jim Walker of Dayton, Ohio. Among them was a 1950 Buick originally chopped and rebuilt in the 1950s by Sam Barris, a legendary California customizer, who used the car himself. Then, "somehow it got to Massachusetts, where it rusted away in a field" until Walker came across it, rescued it, and brought it back to life.[75] Likewise the "Orbitron," a bubble-top, fiberglass show car built in 1964 by one of the custom world's most brilliant minds, Ed Roth, was relocated after more than forty years of rumors and speculation as to its whereabouts. Its long-term hiding spot? "A public sidewalk, in front of a sex shop in Juarez, Mexico," Hot Rod's Pat Ganahl reported in a 2008 article on its recovery, "less than half a mile from the bridge to El Paso." There, "millions of people walked right by it for years," oblivious to the strange car's history and value.[76] Hot Rod, Super Chevy, and Popular Hot Rodding also ran numerous accounts of somewhat less dramatic salvage-yard and other outdoor finds in the 1980s, 1990s, and 2000s, including a 1969 Shelby GT 350 Mustang discovered next to a house in El Paso, a 1967 Nova found in a vacant lot in Southern California, a 1933 Ford woody wagon rescued from the desert in the Four Corners region, a 1965 Chevy II wagon saved from a salvage yard, and a 1965 Mustang Fastback found in a backyard in Austin, Texas.[77]

Stories of this sort were nothing new. As we saw in chapter 2, many a custom, street-rod, and all-out restoration project began with a tattered car—or parts from one—that was found outdoors someplace, often but not always in a wrecking yard. Toward the end of the twentieth century, however, a number of stories involving "backyard finds" and "junkyard

Figure 5. Somewhat puzzling here is the choice of a beautifully restored 1932 Nash for the March 1980 cover of the street-rod-oriented magazine *Street Scene*. Less puzzling is the choice of a stack of junked cars as a backdrop: it's a classic rags-to-riches scene. (*Street Scene*, March 1980, cover. Reproduced courtesy of the National Street Rod Association.)

finds" like these began to assume a slightly different form. Although many continued to turn up in supporting roles in rags-to-riches narratives and how-to features (fig. 5), others instead began to play starring roles in articles and photographic spreads depicting them precisely as they were found: wrecked, cut up, and rusting away.

"Weathered Wheels"

Coverage of dilapidated cars in situ in fields and salvage yards began to appear on a limited basis during the 1970s in the street-rod oriented *Rod and Custom*'s Vintage Tin section. From there it rapidly spread to other periodicals during the 1980s and 1990s; by the end of the 2000s it was ubiquitous. For antique, classic, and special-interest enthusiasts, there were the Weathered Wheels and Wreck of the Week sections in the newspaper *Old Cars Weekly*. For high-performance enthusiasts, there was *Hot Rod*'s Hidden Treasure of the Month. For road-racing and sports-car fans, there was *Classic Motorsports*' Ran When Parked and *Classic and Sports Car*'s Reader Find of the Month. For modern tuner-car enthusiasts, there was *European Car*'s Money Shot, which frequently featured derelict vehicles and parts. For vintage Volkswagen fans, there was for many years the Fertig segment on the last page of *VW Trends*, as well as occasional coverage in *Hot VWs*. For Gen Xers, there was a short-lived magazine in the mid-2000s called *MPH*, which among other lewd and lurid features carried coverage of totaled Porsches, Ferraris, and Lamborghinis in a section titled Wrecked Exotics. For more mainstream readers, there was for many years the PS page in *Road and Track* and the Reader Sightings feature in *Car and Driver*. *Car and Driver*'s annual Top Ten or Ten Best features often focused on wrecked, junked, or otherwise abandoned vehicles as well. The restoration-oriented *Cars and Parts* went much further than the rest, running deep and lavishly illustrated coverage of enthusiast-oriented salvage yards in nearly every issue published since the early 1980s (figs. 6–8). Finally, for those with more than a passing interest in junked automobiles, there were prints of artists' renderings, video tours of well-known old-car yards, photographs of wrecking yards in web-based collections, numerous books on salvage yards and their stocks of rusted relics, and calendars featuring brilliant, full-color images of wrecked and abandoned cars in a variety of settings.[78]

On the surface, the visual content of these features, websites, books, and prints differed little from the "ruins porn" produced by so-called urban explorers: those who trek, cameras in hand, through the ruins of the American Rust Belt and other deindustrialized regions here and abroad, documenting urban blight and industrial decay for distribution on various online forums.[79] Unlike serious scholars of industrial archaeology, however, these explorers and their fans in cyberspace are less interested in the

8. *Its engine bay open to the elements, and its nose on the ground, this '67 Mercury Cougar still may have some of its "nine lives" left.*

9. *A '57 Chrysler Saratoga shows signs of considerable wear and tear, but can still provide its front clip, glass and some sheet metal.*

10. *Body panels are about all that's still salvageable from this '61 Rambler American convertible.*

11. *This '50 Pontiac's body panels and trim could be saved, but its roof and glass cannot. The bumper on the ground does not go with this car.*

12. *Scavenged for parts already, this '60 Ford Galaxie can still yield body panels, glass and trim pieces.*

13. *This '66 Ford has a nice interior and straight sheet metal.*

14. *If you only needed the rear half of a '56 Buick wagon, then this hulk may fit the bill, since its front clip, engine and windshield have already been removed.*

15. *Missing rear tail lights, this '56 DeSoto Firedome Sportsman four-door, despite a damaged roof and doors, still has parts to give.*

Figure 6. A page from a typical salvage-yard feature in *Cars and Parts* in which the author not only documents interesting examples from the yard in question but also notes their condition and the types of parts that each has left to yield. (Eric Kaminsky, "Pearsonville Auto Wrecking & Hubcap Store," *Cars and Parts*, July 1998, p. 48. Reproduced courtesy of Amos Press, Inc.)

Figure 7. From a 2008 *Cars and Parts* feature, a full-page image of a Pontiac rusting in peace in an Illinois salvage yard. The image appeared in full color in the magazine. (Lyle R. Rolfe, "Ace Auto Salvage," *Cars and Parts*, March 2008, p. 59. Reproduced courtesy of Lyle R. Rolfe.)

material reality of the industrial past than they are in its present state of decay. The same was true of at least some of the aforementioned wrecked- or derelict-car features. Like "ruins porn," they often conveyed a sense of resignation—sometimes mixed with an almost voyeuristic delight—as yet another piece of the industrial past slipped away, whether a factory on the East Side of Detroit or the remnants of a car assembled there.

Figure 8. In this magazine-shoot outtake, longtime junkyard sleuth and frequent *Cars and Parts* contributor Phil Skinner captured a 1965 Cadillac hearse with foliage beginning to grow through its open engine compartment. (Reproduced courtesy of Phil Skinner.)

Consider a few examples. In January 1993, as part of its annual Ten Best special, *Car and Driver* ran a series of photographs titled "Ten Best Awful Automotive Relics." Among them were a 1930s Ford buried under miscellaneous debris in a backyard in central Pennsylvania, an old school bus rotting away in a field in central Tennessee, a Corvair behind a rural outbuilding in central Florida, and even a Messerschmitt microcar tucked away on a porch in northern Alabama. Notably, *Car and Driver*'s relics that year also included two examples, a prewar Chevrolet and a postwar Buick, which suffered from a type of postmortem injury common among automobiles left to rot outdoors: damage from trees growing through their bodywork.[80] Over the next several years, *VW Trends* published a number of similar pictures, submitted by readers, of the mortal remains of vintage Volkswagens they encountered here and there. These included the shell of a Microbus, riddled with bullet holes, discovered in a field in Washington State; a badly wrecked, heavily rusted, and partially stripped Beetle in a salvage yard someplace in the Rust Belt; several Beetles and Type IIIs in an overgrown wrecking yard in Northern California; an utterly demolished Squareback in a salvage yard in Southern California; and a Beetle long since left for dead in a wooded area, complete with a large tree rooted between its hood and front bumper.[81] *Hot Rod* ran similar images from time to time in the 1990s and early 2000s before launching a regular feature devoted to them, the reader-submitted Hidden Treasure of the Month, in 2004. Among those chosen to appear in this space in the years that followed were the shell of a 1970 AMC Rebel, decimated by rust in a Massachusetts salvage yard; a 1969 Dodge Charger buried in briars in southern Oregon; a 1979 Dodge Lil' Red Express truck, covered with ivy and slowly returning to earth in eastern Nebraska; and a 1960s Mustang, wedged against a barn by a very large tree, in central Pennsylvania.[82] Other monthly titles ran images of everything from a 1958 Edsel in a roadside field in central Ohio to a largely intact 1959 Chevrolet wagon in a salvage yard in northern Kansas, a 1970s Volkswagen van left to rot in rural Australia, a 1960s Ferrari pulled from a watery grave in Narragansett Bay, a 1925 Bugatti fished from a lake in northern Italy after seven decades under water, and a late-1930s Chevrolet coupe in the ghost town of Bodie, California.[83] *Hot Rod* even dabbled with the "ruins porn" motif directly, publishing a lengthy feature on the abandoned factories of Detroit in 2009 that would have been right at home on many urban-exploration websites.[84]

Other photographic features on wrecked or otherwise out-of-service cars focused less on where they ended up than how they got there. In 1991, *Hot VWs* published several somber images of vintage Volkswagens damaged beyond repair in 1990 after an arsonist started a wildfire near Santa Barbara, California.[85] Other gearhead monthlies occasionally ran similar photos capturing moments of on- or off-road woe as well, including a *Road and Track* image of a Volkswagen camper on its side on a snowy stretch of highway in 1990, a *Hot Rod* snapshot of a Plymouth Road Runner on its roof after a single-car accident in 1999, a *Car Craft* photo of a Buick Sport Wagon engulfed in flames on the side of the road in 2000, and a *Hot Rod* shot of a Chevrolet Malibu drag car taken moments after it was rear-ended—while on its trailer—by a Jeep in 2002.[86] In 2005 a number of monthlies also published images of a multicar pileup at a vintage racing event at the Road America course near Elkhart Lake, Wisconsin. This major mishap involved several early Corvettes, a genuine Shelby Cobra, and about a dozen other vehicles with storied racing histories.[87] Some gearhead periodicals occasionally printed photographs of accidents taken in the more distant past, too. *Cars and Parts* published a reader-submitted image of the aftermath of a 1933 wreck on old U.S. Route 12 in 1998, while *Old Cars Weekly* took to running images of collisions from the 1950s, 1960s, and 1970s in its Wreck of the Week feature in the 2000s.[88]

One could go on and on—the well of possible examples is virtually bottomless. But for the present purposes, the reasons this sort of coverage remained a staple of the enthusiast publications business for so many years are far more telling than any specific case in point. As twisted as it might seem, humor often topped the list. "Clarence lowers his car the hard way," *Car and Driver* quipped in a caption accompanying a Reader Sightings photo of a Pontiac Sunfire crushed by a retaining wall in 1997. No caption was necessary in another installment published the following year, in which the door of a General Motors J-Body car that was absolutely devastated in an accident carried a telling legend: "Sears Driving School."[89] Similarly, *Road and Track's* editors found humor in a photo published on its PS page in 1983 of a man contemplating the aftermath of an accident in which a Ferrari slid off the road and into a grove of trees. Their caption? "'Good tires,' Bob mused, casually lighting a cigarette, 'but certainly not great tires.'" *Road and Track* also found the lighter side of a crash that ended the working life of a subcompact in the 1990s. "I brake for animals," explained a bumper sticker on the car, which was hit from behind. "Others don't," joked the editors.[90] *Road and Track's* staff also had a knack for finding humor in brief tales of woe. In 1982, they ran a story about a Canadian motorist whose car, a 1972 AMC Ambassador, was mistakenly sent to the crusher by a salvage-yard operator while its owner browsed among the wrecks. Referencing the move to smaller cars then in vogue in the United States and Canada, the editors joked that the Ambassador's owner had "become an unwilling member of the movement to compact cars."[91] Cartoons in magazines from across the enthusiast spectrum found humor in nearly every aspect of the gearhead experience as well, including old-car restoration, parts and project hoarding, barn finds, and treasure hunts

Figure 9. Treasure. This George Trosley cartoon appeared in the November 1991 issue of *VW Trends*. (Reproduced courtesy of George Trosley.)

(fig. 9).[92] More sordidly, the monthly Wrecked Exotics feature in the short-lived *MPH* mixed crude and largely unrepeatable humor with a clear disdain for those rich enough to afford Ferraris, Lamborghinis, or Bentleys but too ham-handed to keep them on the road.[93]

Other coverage of wrecked or out-of-service cars appeared in gearhead publications for a more practical reason: a widespread desire to save as many of them as possible. As discussed in the previous chapter, the monthly salvage-yard features in *Cars and Parts* were popular in part because they helped enthusiast-specialty wrecking yards sell parts, systems, and complete cars but also because they helped keep old-car gearheads informed about what was available and where. Its editorials often did the same. David L. Lewis, for example, the author of the magazine's long-running

Ford Country column, ran an image of a Model T rusting away in a field in April of 1990. "Don't let the condition of this relic deceive you," he explained. "Many of its sisters, no less dilapidated, have been restored to mint condition, and are treasured by their owners. Your commentator knows, having a 1921 touring [car] which went through such a metamorphosis."[94] The implication was clear: get out there and save one! More explicit were a series of Salvage Yard Alert notices published in *Cars and Parts* during the 1990s. Each of these pointed to an old-car yard that was either about to close or "go modern," urging readers to save—that is, trek out and buy—the parts and cars within them before it was too late.[95] Many of the magazine's ordinary salvage-yard features in the 1980s, 1990s, and 2000s did the same, especially when the yard in question was on the brink of closing.[96] Readers also chimed in, writing letters to the editor that called on their fellow gearheads to help save parts and cars from this or that endangered yard and to comb their local barns and backyards for treasures yet to be discovered.[97]

Monthly installments of *Hot Rod*'s reader-submitted Hidden Treasure of the Month often made the same points. "Good tin is still out there," *Hot Rod*'s editors explained a month after the feature debuted. Accordingly, their new column was "about sharing your best finds with the rest of the world, but not necessarily revealing their locations. If you've got quality junk with patina, send photos here."[98] Many of those who submitted their "hidden treasures" over the next few years did indeed elect to keep their locations to themselves, but more than a few also urged other rodders to get off their sofas and contribute by making their own discoveries. "My advice to everyone," Duff Gray of Billings, Montana, wrote in a brief letter accompanying a photograph of his 1967 Camaro in as-found condition, "is to keep looking and follow up on rumors about cars for sale—sometimes they're true!"[99] Gearheads partial to other niches—street rods, imports, sports cars—often shared similar photographs, and similar advice, with their favorite monthly titles as well.[100]

As the hunt progressed, some also began to contemplate an important corollary of the search for forgotten antiques, classics, and special-interest cars: the touchy subject of whether, or to what extent, a given find ought to be restored. From the 1920s through the mid-1980s, this was not an issue. Old-car enthusiasts who discovered their projects in barns, used-car lots, and the classifieds nearly always intended to restore them to their former glory. But by the end of the 1980s, some gearheads had taken a page from those who collected other kinds of antiques, like furniture, and had begun to value the patina that a car acquired over the years. In 1988, for example, *Road and Track*'s Peter Egan announced that he planned to keep the worn, scratched, and repeatedly mended aluminum body panels on his Lotus Seven project car rather than replace them with shiny new units, as he originally intended. "What held me back," he wrote, "was the thought of destroying a small piece of history." A twenty-four-year-old "with a lot of racing behind it," the car was covered with revealing scars. "The side panels were slightly creased where someone had bent them back out of the way to reweld the frame," he explained, and "there was a small dent in the flat rear-body panel where the nose of another car had no doubt leaned on the Lotus, hoping to lighten its rear traction in a corner." Egan therefore found himself less than eager to wipe all traces of the car's past out of existence by insisting on a showroom-quality restoration. Other fans of vintage sports cars helped him arrive at the decision. "I never replace anything I can save," one Jaguar enthusiast told him at a meet in Seattle. "You see those old factory inspector's chalk marks on the back of a dash panel and you realize the whole car is full of English ghosts. If you let them escape," Egan's new friend added, "they never come back."[101] Nearly twenty years later, when patina fever was in full swing, *Hot Rod*'s David Freiburger agreed. "Patina lends proof of life," he explained. "It tells a saturated story of age, of history absolutely unrevised. Its unique character of textures can't be bought for any price or fabricated with even a pretense of dignity."[102] More than a dozen photographs of heavily weathered old cars followed, rounding out Freiburger's piece.

From time to time letters, features, and images of wrecked or otherwise less-than-perfect cars served as warnings, too. In a note enclosed with a picture of his 1964 Volkswagen, which was utterly demolished in a T-bone accident in 1993, Mike Singleton of Morgan Hill, California, offered some advice to other enthusiasts. "The cause of this accident was bad brakes," he explained, "so kids, make sure your car stops before it looks cool!"[103] Likewise, *Hot Rod* reader Bob Mosher learned a lesson from an accident that claimed his 1965 Dodge Coronet sedan in 2006. Stored in a building owned by a friend in the fireworks business, the car was totaled in a major blaze ignited by his friend's stock. Mosher's advice, as Steve Magnante reported in a brief article accompanying before-and-after shots of the Coronet, had nothing to do with the relative wisdom of storing a cherished car adjacent to explosives. Instead, it centered on the question of insurance. Although the car was covered for many years at an agreed value of $55,000, it would have cost him $90,000 to build another like it after the fire. "I neglected to increase the coverage as the car's value went up," Mosher told Magnante, so "if you have collector-car insurance, make sure your car gets re-appraised every year so the insurance policy keeps up with its value."[104] Some of the earliest images of wrecked cars that appeared in automotive periodicals involved warnings, too. Back in 1905, the Badger Brass Manufacturing Company ran an advertisement showing the remains of a car following a fatal accident caused by bad lighting—the implication being, of course, that had the driver only taken the time to fit a pair of Badger Brass lamps, he might still be alive, and his car unscathed. Over the next century, similar images were used to sell readers on the need for everything from adequate fire extinguishers and collector-car insurance to safety courses for new teen drivers.[105]

But perhaps the most revealing reason for the frequent appearance of in situ wrecks in enthusiast publications and other media is one that has little to do with making jokes, saving cars, or warning others not to let their own rides end up as heaps of twisted rust. Instead, it centers on the physical and emotional experience of the salvage-yard treasure hunt itself. From time to time this involved an appreciation of the sights, sounds, and even smells associated with wrecking yards. "Prepare your schnozz for the unexpected," Daniel Strohl explained in a brief 2007 *Hemmings* piece on the olfactory dimensions of the junkyard crawl, which might include everything from "the stench of aged cigarette smoke" and "mildew, mold and fungi" to the occasional whiff of four-legged decay. Rotting fabric, he added, might also conjure up "memories of gramma's couch."[106]

Visual cues were apt to trigger moments of broader contemplation (fig. 10). As Bob Stevens of *Cars and Parts* put it back in 1982, "strolling through an auto graveyard is, indeed, an enchanting experience. . . . A long, studied look at a junker can produce visions of former grandeur as one imagines what a particular car looked like when it was new; how it served its owners over the years, and how long it had been discarded among the weeds and mud of a desolate mechanical boneyard." Twenty-five years

Forlorn and Fern-bound

JENNIFER WELLES

BY MARK J. McCOURT HEMMINGS MAGAZINE

The rusted farm registration license plate from 1962 that adorns this 1940 Nash 6-cylinder business coupe offers faint glimpses into its mysterious past life. Sunk to its sills in red dirt atop a ridge that overlooks the west branch of the Neversink River in New York's Catskill Park, the big Nash, camouflaged with rust, surprises hikers by seemingly appearing out of nowhere. With its broken glass and partially caved roof, the interior is beyond hope; misplaced trim, errant bullet holes and rusted floorboards relegate this weary workhorse to fester among the ferns for eternity.

Figure 10. An example of a junked-car feature from *Hemmings*, which appeared in full color in the original magazine. (Mark J. McCourt, "Forlorn and Fern-bound," *Hemmings Motor News*, December 2003, p. 80. Reproduced courtesy of *Hemmings Motor News*.)

later, *Hot Rod*'s Rob Kinnan agreed. "When I see a faded old sedan in a junkyard," he explained, "I imagine how proud the new owner was when he brought it home for the first time—how shiny it was."[107]

Over the intervening years, a number of others chimed in, too. *Car and Driver*'s editors noted in 1988 that "old cars, even in repose, still stimulate the imagination. Their sun-faded colors contrast vividly with . . . memories still bright in the mind. Each [corpse] has a story to tell, and no way to tell it."[108] Ten years later, Frederick Inocencio Collazo, a professional photographer, agreed. "When vintage Volkswagen enthusiasts cross the threshold into this colorful domain of Volkswagen shells," he explained in a *VW Trends* feature on his salvage-yard photography, "they often have thoughts on how 'life' may have been for [them]. . . . 'Where have they been?' 'Who owned them?' and 'How did they get here?'" The following year, Peter Egan of *Road and Track* put it slightly differently. Reflecting on a recent trip he made with a friend to view a private hoard of European cars—or what was left of them—Egan explained that he and his friend were mesmerized by a basket-case Porsche 356B Roadster. "Here's this awful car," he wrote, "and we stand around looking at it from different angles, speechless, as if warming ourselves with the fading rays of heat from the genius of Ferry Porsche or Erwin Komenda. Or perhaps soaking up faint vibes of energy of the stolid German craftsmen who put it together so long ago."[109]

The salvage-yard treasure hunt was more than just a quest for elusive parts and desirable project cars. It was as well a search for meaning and connection, an attempt to catch a glimpse of a material past once taken for granted but now recognized as fleeting.

As many gearheads learned to their dismay, the hunt itself was also fleeting.

Who Moved My Junk?

In 2001, *European Car*'s Kevin Clemens needed a set of bucket seats for a rally-car project, so he set out with his son to look for a set in a salvage yard. This turned out to be a bit more difficult than Clemens anticipated. "The first problem we faced was to find a junkyard," he wrote, for many of the places he used to frequent were by then streamlined "auto parts recyclers" that no longer welcomed wrench-wielding bricoleurs who needed to browse around in order to figure out exactly what they were after. In the end, Clemens and his son found a sympathetic yard and were able, after a bit of scrounging, to find a pair of seats that fit the bill. "But, junkyards like the one we visited are quickly vanishing," he lamented, glad that his young son was able to participate in a genuine boneyard trawl but distressed that future generations might not be able to do so.[110] Numerous features from the 1980s, 1990s, and 2000s about particular salvage yards conveyed similar sentiments—"Old-Car Collectors Must Play 'Beat the Crusher,'" "Endangered Junkyard," "Transporters Find Sanctuary," "Salvage Yard Alert"[111]—as did the subtext of many of the stories of automotive treasure published during those same years. For by then, the laid-back, open-air, browse-at-will salvage yard beloved by enthusiasts was an endangered species in its own right. So too were "de facto junkyards," the random gatherings of derelict cars in rural and suburban areas that were often the source of cherished automotive discoveries.[112]

This was due in part to shifts within the salvage industry. The nature of the wrecking business itself had changed during the 1970s, 1980s, and 1990s in ways that would ultimately jeopardize Saturday-afternoon strolls through weed-covered relics. Among others, these included the advent of crushing and shredding technologies and the adoption of a streamlined business model favoring rapid turnover. More broadly, the salvage industry faced mounting legal obstacles. The regulatory environment grew less and less welcoming to secondary-materials firms of all sorts during this period, for reasons ranging from air pollution and ground and soil contamination to widespread angst over aesthetics and neighboring property values. Similar concerns placed older cars themselves in the crosshairs, as federal, state, and local initiatives designed to eliminate them from daily use gained traction. Businesses fought back and occasionally won reprieves, as did individuals with parts- and project-cars in their driveways. But as often as not, junkyards closed, older vehicles were forced out of sight (and sometimes nudged out of use), and enthusiasts felt the parameters of the old-car hobby shifting all around them.

The final chapters of this book examine these legal and regulatory developments, focusing on aesthetics, property values, and zoning in chapter 5 and old-car scrappage in chapter 6, as well as the individual and collective responses of gearheads everywhere. For by the 1990s, "save the old cars" was no longer just a slogan to inspire the hunt for barn finds and salvage-yard treasure. Instead it was a broader imperative, one reflecting a determined fight to preserve and maintain the automotive past in all its forms.

5

Not in My Neighbor's Backyard, Either

Zoning and Eyesore Ordinances, 1965–2010

In the spring of 1999, Daniel Groff snapped. For more than twenty years the Elizabeth Township, Pennsylvania, man had been involved in a dispute with local officials over the condition of his property. Groff worked as a long-haul trucker, agricultural laborer, and mechanic, and over the years, old cars, trucks, heavy equipment, and parts of all sorts had accumulated on his land. For Groff, this was an essential repository of spares that enabled him to scratch out a living. But in the town's view, it was an "illegal junkyard," and Groff an irresponsible property owner. Toward the end of 1998, after countless hearings, injunctions, and appeals, the town gained the legal upper hand and notified him that it planned to hire a contractor to clear his land. But when the contractor arrived the following March, Groff refused to cooperate. Armed with a shotgun, he used a front-end loader to charge at the contractor's equipment, disabling it by pushing it from its trailer. He then retreated to a defensive position, loader idling and shotgun in lap. The ensuing standoff lasted until the fuel in his loader ran out, and as the police prepared to move in, Groff turned the gun on himself. Shortly thereafter his land was cleared, and, adding posthumous insult to injury, his grieving widow then received a bill from the township for the contractor's services.[1]

Apart from its grisly ending, Groff's case was not unusual. Similar, though typically nonviolent disputes between municipalities and property owners over the accumulation of "junk"—derelict cars and car parts, in particular—have occurred with increasing frequency since the 1970s. Likewise, disputes pitting local officials and abutting property owners against established automobile graveyards and scrap-metal businesses have been on the rise since the 1960s, dramatically so since the 1980s. Lady Bird Johnson's grand initiative, the federal Highway Beautification Act of 1965, is at least in part to blame.[2] Few would dispute its enduring benefits. American highways are certainly cleaner and sometimes altogether billboard-free, and the salvage yards and scrap-metal businesses along them are normally screened from view. One cannot help but wonder, of course, whether the most common form of screening—razor or barbed wire–topped barriers made from sheets of scrap metal—actually represents an aesthetic improvement over visible stacks of Impalas and K-Cars. Nevertheless, daily commutes and family road trips have been more or less free of the visual assault of litter, unrestricted advertising, and the remnants of yesterday's cars for fifty years—surely this warrants a tip of the hat to the Johnson administration. However, Lady Bird's Bill also spawned a broader cultural shift less worthy of praise. For in its wake, the bill's broad emphasis on "beautification" gradually gave rise to a far more radical wave of antijunkyard, anti–scrap yard, and, for lack of a better phrase, anti-ratty-looking-car NIMBYism over the last five decades.[3]

Simply put, community leaders and homeowners across the United States decided during the 1970s, 1980s, and especially the 1990s and 2000s that Johnson's bill did not go far enough. Most innocently, they ratcheted up the effort to eliminate abandoned vehicles from vacant lots and public streets. But they also targeted long-established urban, suburban, and even rural secondary-materials businesses, chiefly through bylaws, licensing requirements, and zoning restrictions. They (ab)used the power of eminent domain to eviscerate entire industrial districts containing salvage yards and other unpleasant businesses. They went after their gearhead neighbors, too, using municipal bylaws to redefine residential property with unsightly project vehicles on the premises as "illegal junkyards," subject to the same zoning, licensing, and other regulations as profit-seeking entities. Perhaps most egregiously—and with increasing frequency in the 1990s and 2000s—real-estate developers on the rural-suburban fringe subdivided tracts of land adjacent to long-established salvage yards and then worked with their new homeowners to pressure town officials into forcing the abutting eyesores out.

Perhaps these sorts of conflicts over land use, property values, and beautification amount to nothing more than an American farce, complete

with country bumpkins, blue-collar ruffians, snooty yuppies, and unscru-
pulous politicians. Or perhaps they merely represent another chapter in
humanity's long search for the "ultimate sink," a place to dump those things
we deem unsightly or unclean—in this case, out-of-service combines and
clapped-out cars. But if we dig a little deeper, and if we really listen to what
the actors in these stories have to say, we might not be so eager to brush
them aside. For at their core, the many disputes over neighbors' cars, neighbors' yards, and neighboring junkyards that have played out over the last
fifty years reflect not only here-and-now concerns about property values
and eyesore mitigation but also a lingering divide within American culture
over the meaning, value, and broader implications of a way of life geared
toward obsolescence. Put another way, these stories present a clash of
worldviews centered broadly on the relative merits of the old and the new.

For most Americans, the automobile and obsolescence have long gone
hand in hand. With a life expectancy of seven to eight years, the typical car
was meant to be produced, sold, driven, resold once or twice, and then retired to a salvage yard. After several weeks, months, or years, depending
on the yard, what remained of the car would then be sold wholesale to a
scrap yard. There it would be processed for delivery and sale to the steel
industry, where it would be smelted into new materials for the production, among other things, of new vehicles. According to this logic of obsolescence, the older a car became, the more it was an eyesore, a relic of a less-
sophisticated automotive past. Likewise, automobile salvage yards and
scrap-metal yards were at best unsightly links in this chain, businesses of
a type best relegated to the margins of the community (and, if possible, beyond). However, a sizable minority of motorists—chiefly antique, classic,
and special-interest enthusiasts, as well as hot rodders, street rodders, and
others—actually *valued* older cars, avidly restoring, driving, and collecting
what others discarded as obsolete. What's more, they cherished junkyard
businesses and out-of-service project and parts cars not as necessary evils
but as precious resources for the maintenance and refurbishment of the vehicles they treasured. Similar in practice if not in circumstances or mind-set
were the many people on the margins of the socioeconomic landscape, those
who eked out a living fixing this and that with parts scavenged from infor-
mal junkyards on their property—folks like Daniel Groff.[4]

Understandably, scrap-yard owners tended to disagree with the notion
that their businesses were eyesores. Instead, to them their yards were the

source of an honest living made by turning yesterday's scrap into tomorrow's steel, a living firmly rooted in what has long been a mainstream value and one of the prime movers of middle-class prosperity: planned obsolescence. On the other hand, much of the profitability of the automotive salvage business has long derived from recovering major parts and systems from wrecked vehicles and reselling them for use in the repair of other cars, prolonging their useful lives. Consequently, in the eyes of a middle America convinced of the fundamental goodness of the new and backwardness of the old, these yards were guilty not only of being unsightly but also of undermining the logic of obsolescence by making it possible for older machines to stay on the road. Were it not for salvage inventories, that is, the neighbors' primer-spotted, metallic-pea-green, Nixon-era Buick would long ago have succumbed and given way to something newer, shinier, and more in step with the times. After all, what's good for General Motors is good for America—and for American property values. Fortunately for middle-class sensibilities, with each passing year another crop of Groffs was shunted aside, another cluster of junkyards was cleared, and another group of enthusiasts in Anytown, U.S.A., were forced to sell their project cars to satisfy their neighbors and councilmen. In other words, whether or not any of us approve of the activities of salvage businesses and old-car gearheads or their presence in our towns, the fact of the matter is that they have long been on the retreat. So too have their routines centered on repair, restoration, and creative reuse.

Therein lies the subject of this chapter. Focusing on urban, suburban, and rural land-use conflicts between municipal authorities and salvage businesses, de facto junkyard properties, and automobile enthusiasts, the pages that follow explore the ways in which five decades of beautification actually worked to reinforce a throwaway mentality. Paradoxically, that is, we litter less now than we did in 1965, but we waste a great deal more.

The Accelerated Marginalization of the Automotive Salvage Business

The origins of Lady Bird's Bill have been carefully addressed at length by others, notably Carl Zimring and Tom McCarthy, but for our purposes what matters is that during the 1950s and 1960s, three trends gradually converged. The first, which emerged as early as the 1920s and was endemic to urban areas four decades later, was the abandoned-car problem. Whether stolen and stripped, broken beyond their owners' skills or finances to repair, or simply no longer wanted, cars left to rot on city streets or in vacant lots represented a serious problem. High scrap-metal prices kept this blight in check during the 1940s and early 1950s, but by the middle of the 1960s tens of thousands of cars were being abandoned each year in New York City alone. The numbers themselves were staggering, but the crux of the problem lay in the rapidity with which vandalism and parts theft could turn such a car—even one in quite presentable condition when abandoned—into a burned-out eyesore: twenty-four to seventy-two hours.[5]

The second trend was the accelerating rate at which middle-class Americans traded in their old cars for new ones during the 1950s and 1960s. This, in tandem with record-setting total registration figures, meant that the volume of unwanted cars that ended up in salvage yards each year rapidly rose. Coupled with low scrap prices, this resulted in a growing number of overcapacity salvage yards that often had no choice but to store their wrecks in unsightly stacks. By the middle of the 1960s, middle America was ready to see these towering eyesores toppled.

The third trend was relatively straightforward and centered on road-side blight of a different sort. Gaudy billboard advertising and litter of all kinds spread in tandem with the interstate highways. This, together with the urban plague of abandoned cars and the rural and suburban phenomenon of the overflowing junkyard, inspired the first lady to push for action. Thus was born the federal Highway Beautification Act. But, following an initial burst of activity and enthusiasm in the late 1960s and early 1970s, much of the federal effort to regulate salvage properties shifted away from aesthetic concerns and toward more fundamental challenges like soil and groundwater pollution. By the early 1980s, an era of Superfund cleanups and criminal environmental prosecutions, beautification had all but vanished from the agenda in Washington, D.C.

The opposite was true at the local level, where efforts to clean up salvage properties on aesthetic grounds gained speed. Many town councils and county boards shuffled their eyesore-mitigation ordinances during the 1970s, reinvigorating the push for solid walls or other screening measures. Others tightened their licensing requirements, making annual renewals less automatic for extant yards and licenses for new entrants into the salvage business virtually unobtainable. A number of others made use of new or amended zoning statutes to better manage the physical place-ment of these businesses.[6] Finally, numerous urban renewal and district redevelopment efforts proceeded apace in the 1970s, 1980s, and beyond. Sometimes these projects reflected a genuine desire to transform rough industrial districts into parks and other civic facilities;[7] sometimes they involved little more than an attempt to boost tax revenues by replacing marginal businesses like salvage yards with more upscale restaurants and trendy mixed-use zones.[8] Either way, many scrap and salvage businesses were forced to either close their doors entirely or move their operations to outlying areas.

But by far the greatest local beautification-related threat to automobile salvage businesses, especially since the mid-1980s, has been sprawl—more precisely, the sprawling of residential developments into areas on the "rurban fringe."[9] Robert Bruegmann and others have described this as "ex-urban" rather than traditional "suburban" growth, both because of its dis-tance from urban centers and because of its tendency toward larger lots and stately homes.[10] For our purposes, what matters is that new construc-tion in rurban or exurban areas during the late 1980s, 1990s, and 2000s often pitted the owners of long-established rural salvage yards against their newcomer neighbors.

The resulting land-use conflicts were fierce, but rarely did they neatly fit the NIMBY mold. Strictly speaking, "not in my backyard" is a state-ment of *defensive opposition*. Activists in established neighborhoods often cry "NIMBY" when plans to raze an abutting forest for a superhighway are announced or when an established community is chosen by the powers that be to host a new incinerator, pipeline, or any of a number of other facilities described in the secondary literature as "locally unwanted land uses" (LULUs). The key here is that "not in my backyard" implies the exis-tence of a literal or metaphorical backyard *prior* to the emergence of a given LULU plan.[11] On the other hand, salvage-yard conflicts on the rurban fringe much more commonly involved precisely the opposite scenario: long-established secondary-materials businesses that became LULUs only when a sufficient number of *new* backyards were deliberately sited adjacent to them.

Planning and land-use scholars have occasionally described a similar phenomenon involving the construction of new homes in rural areas whose owners then proceed to complain about the sounds and smells of neighboring farms. Likewise, others have described the closure of automobile racetracks following the adjacent construction of new homes, attempts on the part of newly built neighborhoods to crowd out neighboring airstrips, and similar scenarios.[12] But what to call the state of mind behind these stories? NIMBY by itself is ill-suited, as are NIABY ("not in anybody's backyard," used to describe facilities of such a dangerous nature that they ought never be built at all, no matter where) and PIBPY ("place in black people's yards," used by activists in the environmental justice movement to protest the all-too-common practice of locating new LULUs in majority African American districts). "Rural gentrification" does not quite fit, either. It hints at the overall effect of the process of rurban cleansing at the behest of wealthy newcomers, but it does not speak to the legal mechanisms involved, nor to the operative mind-set. Thus, for lack of a suitable alternative, this chapter describes the phenomenon as "ex post facto NIMBYism." Though it doesn't exactly roll off the tongue, "ex post facto NIMBYism" does more accurately describe what has so often happened since the middle of the 1980s.

A word of clarification is in order regarding the mechanics of ex post facto NIMBYism. Zoning changes—as well as the establishment of altogether new zoning rules, both in freshly incorporated areas and in long-established towns that previously had no zoning regulations—usually allow preexisting land uses to continue indefinitely, a practice known as "grandfathering." This affords the owner of a property that becomes nonconforming upon a zoning change some protection against what would otherwise amount to retroactive prosecution. When a village incorporates and adopts a zoning rule prohibiting barns on nonagricultural property, for example, they typically do not require that preexisting barns be razed. The same has long applied to extant salvage yards: most of the time, grandfathering protects them when zoning rules are established or changed. New salvage yards cannot locate where prohibited by new rules, but extant yards are normally allowed to remain.

But even when this is the case—even when zoning regulations allow "nonconforming" uses to continue—incorporated areas remain free to pass additional laws regulating the behavior of preexisting businesses in other ways. This gives officials the legal leeway they require if they wish to force a nonconforming type of business out. "The advantage to a general bylaw on this issue in addition to, or in place of an amendment to the zoning bylaw," explained the legal counsel to the town of Uxbridge, Massachusetts, in 1989, as its selectmen debated a new set of rules for extant salvage yards within their jurisdiction in light of recent housing construction, "is that there is no 'grandfather' protection from general bylaws."[13] Accordingly, if a salvage yard established in 1945 in Uxbridge (or anywhere else, for that matter) was allowed to continue to operate after its environs were zoned residential in 1985, say, town officials could nevertheless still respond to ex post facto NIMBYist complaints voiced in 1995 by passing new rules regulating all businesses of its type. If properly drafted, these new requirements could easily make it impossible for our hypothetical yard to continue to operate profitably, and eventually its owners would have no choice but to throw in the towel. Consider a few examples.

In the mid-1980s, a small cluster of homes was built on a cul-de-sac in Kewaskum, Wisconsin, on a site adjacent to an automobile graveyard. That business, Don's Salvage Yard, had been in operation there for more than three decades. Several years before the neighboring homes were built, Kewaskum passed a new zoning ordinance stipulating that the entire area surrounding Don's was to be used for residential rather than commercial or industrial purposes. But because it was established long before the ordinance took effect, the operation of Don's Salvage qualified as a legal, nonconforming use of the land. Nevertheless, complaints from homeowners regarding the yard's appearance began to surface during the late 1980s and were frequent by the early 2000s. At this point, the town invoked a set of bylaws requiring all salvage businesses, even those located far from major roads, like Don's, to shield their operations from view. In 2004 the yard's owners, Donald and David Stern, agreed to comply with this requirement, but residents and officials remained dissatisfied with the condition of the property and the speed at which the Sterns were working to complete their solid fencing. Three years and more than one hundred citations later, the town filed suit against Don's Salvage, seeking a court order to cease their operations so that proceedings against them for municipal offenses could commence. The Sterns were guilty not only of creating a public nuisance, the town charged, but also of operating an "illegal junkyard" because they never obtained an official Kewaskum permit for their nonconforming use of zoned land. In addition, in spite of a clean bill of health from the Wisconsin Department of Natural Resources, the town charged that the yard was a menace to the environment. Though the Sterns and the authorities arrived at a temporary settlement in 2007, the writing was on the wall: Don's Salvage Yard was no longer welcome in Kewaskum.[14]

Litzkow Auto Salvage in Waterloo, Iowa, faced a comparable scenario. It too was established prior to the adoption of local zoning ordinances, which in this case happened in 1969. Therefore it too operated as a preexisting, nonconforming business. But pressure from residents of newly built homes to the north of the property prompted town officials to seek a means of ending the yard's long tenure in the early 1990s. Their first tack was to file suit against the yard, challenging its right to exist at all on land zoned for agricultural purposes. This fell flat in 1993, when on appeal the Iowa Supreme Court ruled that Litzkow Auto was indeed entitled to continue to operate as a grandfathered business. Eleven years later, Waterloo tried again as part of a broader, town-wide crackdown on junkyard-ordinance violations. Several yards were cited in this effort, including Litzkow Auto, for having failed to erect solid fences around their properties to shield them from view. However, Joseph and Elizabeth Litzkow had been working since the middle of the 1990s to bring their business into compliance with this shielding requirement, only to be rebuffed each time by the very same officials who were now citing them for refusing to comply: the Waterloo Board of Adjustment. The problem was that Litzkow Auto, situated on both sides of a road, was built on a one-hundred-year floodplain. Therefore it was illegal under federal and state law for the Litzkows to build a solid fence around the two halves of their operation. To bring their property into compliance, the Litzkows proposed to fill the land to the north of the road in order to raise it above the floodplain, enabling them to build a solid fence there legally. Then they planned to move their entire operation to this larger, fenced-in, north-side plot. Both the U.S. Army Corps of Engineers and the Iowa Department of Natural Resources gave their blessings to the plan, but in 1997 the Waterloo Board of Adjustment rejected it. Thus, when the Litzkows were cited by town officials in 2004, it was for failing to do precisely what those same officials had prevented them from doing seven years earlier: build a solid fence. In 2007, the dispute resumed when the Litzkows again failed to obtain the town's permission to carry out their fill-and-fence plan, largely because the residents of the neighborhood abutting the north side of the yard did not want to see the operation "expand." Three years later the town obtained a block of redevelopment money and used it to pay the Litzkows to move their business to a less objectionable location farther from residential areas. This confirms what the balance of the story suggests: Waterloo officials were less interested in allowing the Litzkows to clean up their property and bring it into compliance with applicable shielding requirements than they were in finding a way to remove the operation from its latecomer neighbors' lines of sight for good.[15]

Similar stories abound. In Hernando County, Florida, there was for example what can only be described as the saga of the O.K. Corral. Also known as Hubcap City, the O.K. Corral was a general salvage business whose aging owner struggled for years to keep up with the ever-changing rules and regulations governing his type of operation that were passed for the sake of new commercial and residential developments in the area. Though it had its share of supporters—including at least one who wondered whether it was really necessary to force this particular business to clean up its act in order to protect the inflated property values of its newly built abutters, mostly tacky strip malls and shoddily built subdivisions— the O.K. Corral closed its doors for good in 2002.[16] There was also the case of Western Hills Auto Parts in Burrillville, Rhode Island, a long-established business protected by grandfathering from a number of zoning and other ordinances passed in more recent times. Town officials succeeded in using the occasion of the sale of Western Hills to a new owner in 1998 to reset the clock on the yard's grandfathered tenure, enabling them to enforce all of the rules with which its previous owner did not have to comply. Denied a town license to continue operations in 2000 after failing to bring his business into compliance, the yard's new owner was forced to close down after an appeal to state authorities brought no relief.[17] Also germane was the case of Harmon Auto Wrecking in Metamora, Illinois. After more than forty years in the same location, a handful of new neighbors complained about the site in the 1990s, prompting an investigation into the yard's compliance with state shielding requirements. As the legal fees and the costs of bringing the yard up to code mounted, its owner threw in the towel, liquidating his inventory and closing for good.[18] Also relevant was the case of a small salvage yard run by Andrew Zielinski of Oak Creek, Wisconsin. In spite of its legal entitlement to grandfathered status—it opened several years before the town of Oak Creek was incorporated, and nearly fifty years before any zoning rules for its part of town were drafted—Zielinski's yard was forcibly closed in 2009 when town officials refused, under pressure from the owners of new homes in the vicinity, to grant a nonconforming license. Were it not for a last-minute lawsuit filed by his attorneys, the town would have cleared the property and charged Zielinski for the pleasure of its services.[19]

Counterexamples cropped up from time to time as well, often involving short-lived or symbolic victories. There was the case of Gene Crandall, an upstate New York salvage-yard owner who gave the finger to his neighbors and town officials in 2002 after a lengthy battle over property shielding. He did so by exploiting a loophole in the town's ordinances: nowhere did its salvage-yard requirements state precisely what a solid shielding fence could or could not be made from, so Crandall built his out of hideous crushed-car cubes. Unfortunately for him, the town was not amused, and its attorneys responded to his provocation by denying his bid to renew his business license.[20] Meanwhile, in Spotswood, New Jersey, the owner of Giancola Motor Cars also won a short-lived victory in 2002. Responding to complaints from abutting subdivisions built long after Fernando Giancola opened his salvage and used-car business, the town revoked his license that February on the grounds that his shop was in an area zoned for residential use. Later that year, however, a state court ruled in Giancola's favor, ordering the town to restore his license. "I'm in the wrong zone, they say it is a residential," Giancola told the *Home News Tribune* of East Brunswick later that year, but "we are a pre-existing nonconforming use. We were there before the houses were there." Nevertheless, Spotswood went on to issue another citation in 2003, and Giancola, drowning in red tape, ultimately accepted a less-than-favorable settlement.[21]

On the other hand, the owners of Weekly's Auto Parts and Swangler Auto Wrecking, two salvage yards in Grand Forks, North Dakota, managed to secure a more lasting victory. In the mid-2000s, they faced opposition from some members of their city council when their licenses came up for renewal. Concerned about the overall appearance of the stretch of road where they were located—and the subsequent effect of this appearance on the town's ability to dazzle visitors and attract new investment—these officials hoped to use the licensing process to force the yards to tidy up a bit. Chief among their allegations was that Weekly's and Swangler were guilty of storing inoperable cars in plain view, outside their solid fences. According to the head of the Grand Forks Inspections Department, however, most of the vehicles in question were in fact in operable condition, and those that weren't actually weren't the responsibility of the yards. "People drop off cars and just abandon them out there," she told the *Grand Forks Herald* in 2004. The paper went on to explain that these inoperable cars were often "parked just outside the junkyards' property line. Though the obvious intent is for the junkyards to claim the cars," it continued, "the state requires a waiting period before the cars are listed as abandoned property" and thus fair game for Weekly's and Swangler. Though it took some doing, not to mention more than a year of legal wrangling as well as testimony from supportive officials, the yards secured the backing of a majority of council members and were granted their renewals.[22] Yards in several other states managed to win drawn-out battles with town officials and ex post facto NIMBYists in the 1990s and 2000s as well.[23] In addition, more than a few salvage-yard owners in rurban areas actually welcomed the encroachment of new homes as an opportunity to sell out of the business and retire. Some of their yards became golf courses, others shopping malls or more new homes.[24]

One could go on and on, but let us briefly turn instead to what these stories as a group suggest, outcomes notwithstanding. Simply put, salvage yards, widely viewed as disagreeable neighbors, long required both a considerable amount of space in which to carry out their operations and relative proximity to built-up areas from which to draw in customers and source their stocks.[25] As a result, salvage businesses tended to self-locate on the margins of settled areas, where immediate neighbors were few, land cheap, legal requirements relatively lax, and both unwanted cars and used-parts customers in reasonable supply. But over time, suburban and exurban growth brought genteel development into these formerly marginal areas. And as this happened, extant salvage businesses faced an uphill fight.

The Eradication of De Facto Junkyards

Ex post facto NIMBYism has been a problem for salvage businesses operating on the rurban fringe for several decades. But it has always been an even greater problem for an altogether different group, those who have simply accumulated derelict cars, scrap metal, and other miscellaneous items on their properties. Sometimes this occurred in residential areas, often beginning with a clapped-out car or two, or perhaps an old refrigerator on the front porch, before spiraling out of control. Sometimes it happened in commercial districts as well—at auto body shops, for example, that amassed a few too many wrecks on their premises. But it occurred most frequently in rural districts, where the rules regarding inoperable vehicle storage, the accumulation of "junk," and general property-maintenance bylaws have long been weak, nonexistent, or infrequently enforced. During the 1980s, 1990s, and 2000s, however, many rural districts across the United States were absorbed into the rurban fringe as suburbia and exurbia expanded. And as this happened, land-use conflicts involving these unsightly properties became more common.

Whether rural, exurban, suburban, or urban, the powers that be tended to describe these properties as "illegal junkyards" largely so that they could rein them in with statutes applicable to profit-seeking salvage enterprises. In other words, by declaring unsightly, junk-filled yards "illegal junkyards" rather than simply "derelict properties," they were often able to bring harsher penalties to bear on those who owned them, facilitating timelier cleanups. But only rarely have those accused of operating "illegal junkyards" actually been guilty of running profit-seeking businesses on the sly. Instead, most simply accumulated too much stuff for their neighbors—long-established and newcomers alike—to bear. For this reason, I prefer to call these properties "de facto junkyards." This is a more accurate designation than "illegal junkyards," because although it does not try to sanitize the nature of such properties by calling them something other than what they appear to be—junk-filled lots—it also does not make the mistake of conflating haphazardly out-of-control properties with deliberately assembled, profit-seeking salvage entities.

De facto junkyards that formed in extant urban, suburban, and exurban areas are more or less impossible to defend. Consider Southern California's Tom Merkel, whom we met in chapter two. When he began to buy up old and unloved cars and store them on the streets of Santa Barbara in 1973, amassing a collection of derelict vehicles numbering in the hundreds by the early 1980s, who could fault the authorities for forcing him to clean up his unwelcome mess?[26] Or when a Luzerne, Pennsylvania, man began to use his suburban backyard to store scrap metal and automobile parts and his front yard and the street to repair cars and trucks in the late 1990s, who could fault his neighbors for complaining to the city, or the city for taking action?[27] Or when a Holland, Massachusetts, man was cited time and again during the early 2000s for hoarding unregistered cars and other inoperable equipment on his residential property, who could fault his neighbors for their mounting frustration?[28] Or when a Citrus County, Florida, man began to amass on his lawn a number of beat-up cars as well as appliances and other items thrown out by his neighbors in an attempt to repair and resell them in the early 1990s—an example of an actual profit-seeking "illegal junkyard"—who could fault the authorities for attempting to shut him down?[29]

Somewhat less clear-cut were urban and small-town cases involving auto repair facilities, body shops, and towing companies. Because they often centered on a judgment call regarding when exactly it was accurate (and therefore legally actionable) to say that a given business licensed for one type of activity—auto repair, for example—had morphed into an illegal salvage operation, it would be an understatement to say that these cases have rarely dwelled in a black-and-white realm. After all, how many inoperable customer cars awaiting repair are permissible on the grounds of a shop before its owners might legitimately be accused of amassing an "illegal junkyard"? How many abandoned vehicles can a towing company store in its lot before justifiably becoming the target of illegal-junkyard accusations? How many wrecks—more accurately, "duplicates" or "parts cars"—can a body or restoration shop reasonably accumulate before its neighbors earn the right to cry foul? In the litany of cases of this sort that have surfaced over the last two decades, easy answers have been few and far between.[30]

De facto junkyard cases on the rurban fringe, particularly those involving ex post facto NIMBYism, are an altogether different matter. As a general rule of thumb, few would ever want to see any sort of junkyard materialize in their neighbors' yards, in much the same way that few would ever wish to see brand-new scrap yards, garbage incinerators, or landfill operations set up shop next door. But when a group of people knowingly purchase new homes next to an unsightly property, and when as a result of these newcomers' efforts the authorities proceed to identify the property as an "illegal junkyard" and subsequently force it out through newly minted eyesore bylaws or through extant salvage-business regulations, one cannot help but say that something is amiss. The same is true when town or county officials claim to have "discovered" a de facto junkyard and act to forcibly remove it, even though it has been around for years and is either in an isolated spot far from built-up areas or has generated no complaints from its neighbors. This is especially true when the property in question turns out not to be a de facto junkyard at all. Consider a few examples.

In 1961, the stables on Walter Calvin Arnold's farm in Prince George's County, Maryland, burned, leaving him without a place to store his hay and fertilizers. Rather than rebuild the stables, he decided to replace them with a handful of old milk- and laundry-delivery vans he purchased on the cheap from a salvage yard. Arnold later argued that because the vans were less airtight than the comparable (and much more expensive) replacement storage buildings available at the time of the fire, the vans were actually superior for his purposes, since a limited amount of circulation helps prevent mildew. And indeed, for about a dozen years the vans did serve him well in this capacity. During those same years, however, suburban D.C. rapidly expanded in his direction, and none of his new neighbors liked the sight of the vans. By 1973, the newcomers had sufficient strength in numbers to persuade county officials to take a look, and when they did, they decided to issue Arnold the first of many complaints for maintaining an "illegal junkyard." Nine years of citations, lawsuits, countersuits, and appeals ensued, and in the end the county won. In 1982, Arnold was ordered to dispose of the vans, and he was further enjoined from purchasing and storing inoperable vehicles of any type. What irked him most of all, he told one reporter at the time of the clearance order, was the "illegal junkyard" label. "They're gonna throw the book at me," he lamented, for doing something that he never tried to do: operate an unlicensed automotive salvage business.[31]

Frank Stickles of the Columbus, Ohio, area had an even more unpleasant experience with abutting newcomers. Stickles, who restored vintage cars, trucks, and buses as a hobby, moved his large collection of several hundred parts cars and restoration projects three times during the 1960s, 1970s, and 1980s in order to escape the region's suburban and exurban sprawl—and the complaints about the condition of his property which inevitably accompanied it. He eventually settled in Jackson Township on the southern edge of the Columbus area, confident that he had at last secured a home sufficiently far from the urban core to rid himself once and for all of the specter of new developments and ex post facto NIMBYism. By the middle of the 1990s, however, the region's sprawl caught up to him once more, and this time there was no escape. Responding to complaints from the residents of a new development nearby, the township served a notice ordering him to dispose of his collection on the grounds that he was guilty of operating an unlicensed junkyard. Though he occasionally sold the cars he restored, Stickles did not operate a salvage business of any sort, nor a profit-seeking restoration firm. Nevertheless, the penalties piled up, amounting to nearly $140,000 by the end of 1998. Unable to afford the fines—or continue the fight in court, or simply move away again—Stickles folded, and the township seized his acreage and vehicles and auctioned them off in early 1999.[32]

Finally, consider the case of Eugene Mixon of Fairfax County, Virginia. For more than thirty years, Mixon and his wife lived in peace with their neighbors on a quarter-acre plot. A solid fence around their backyard made it impossible for passersby or any of their neighbors to see the half-dozen or so old cars that Mixon parked there. None of them were fully restored, but they all ran and were all legally registered. Nevertheless, when county officials discovered them, they declared the Mixons' property an "illegal junkyard" and began proceedings against the couple. Tragically, these proceedings actually *created* a great deal of strife in a neighborhood where very little had previously existed.

It all began in the summer of 2002, when one of their neighbors complained to county officials about a large two-story shed the Mixons built in their backyard, as well as the height of their fence. Apparently, the neighbor was denied a permit for similar structures and was justifiably annoyed. But when the zoning officials arrived to inspect the shed, Mixon, an employee of the Justice Department, denied their request to enter the property. Undeterred, they proceeded to borrow a county helicopter and conduct a visual flyover of the property. This alerted them both to Mixon's backyard car collection and to a pile of construction materials adjacent to his house; suddenly, the unpermitted shed was small potatoes. On the basis of the construction materials and the relatively large number of cars they spotted from the air, the officials literally slapped an "illegal junkyard" label on the property, erecting a large notice on the Mixons' front lawn declaring that the county planned a hearing on their junkyard-zoning violations. When this sign went up, the neighborhood fractured. Some defended the Mixons, while others bought into the fiction that they were indeed guilty of running a shady junkyard business of some sort. Never mind the fact that all of Mixon's cars were legally registered and fully operable—and thus, in the eyes of the State of Virginia, categorically *not* junk. Never mind, either, that none of them were visible to any of his neighbors. Never mind as well that the construction materials in his yard were there because he was working on the weekends to maintain his home by repairing its roof. Never mind, too, that the *Washington Times* came to the Mixons' defense in an editorial denouncing the waste of tax-payer dollars represented by the flyover itself as well as the trumped-up proceedings. In spite of all of this, in August 2002 the county board agreed with the assessment of its zoning officers and the fears expressed by newly frenzied neighbors: the Mixons were guilty of operating an "illegal junk-yard" in their backyard, and they were therefore subject to legal sanctions until the property was cleaned up to the satisfaction of the board.[33]

Though apparently extraordinary on a number of counts—the helicopter flyover, the provocative prehearing notice, the irrational NIMBY senti-ments the notice itself generated, the willful neglect of critical facts and explanatory details on the part of county officials—the troubles Eugene Mixon and his wife faced were actually rather ordinary in three key ways. First and foremost, quite apart from the fact that nothing in their back-yard was actually illegal, except perhaps the shed, theirs was a case involving an "illegal junkyard," just like the aforementioned examples of Walter Calvin Arnold and Frank Stickles, not to mention countless oth-ers.[34] Second, also like Stickles and Arnold, much of the NIMBY senti-ment that helped propel the authorities' case was newly minted. Stickles and Arnold faced actual newcomers and ex post facto NIMBYism, of course, whereas the Mixons simply faced new fears among longtime neigh-bors. Nevertheless, all three cases hinged on new complaints about old practices. Third, and perhaps of greatest significance for the subject at hand, the "illegal junkyard" troubles Eugene Mixon and his wife faced were the result of a fundamental difference of opinion regarding the nature of the vehicles parked in their backyard. To the Mixons, they were cherished vintage cars. But to the authorities, they were junk—legally registered and operable junk, perhaps, but junk all the same. This is a common difference of opinion regarding older or unkempt cars, and it is practically super-abundant when it comes to property and land-use disputes involving auto-mobile enthusiasts.

Gearheads, Parts Cars, and the Long Arm of the Law

"Automobile enthusiast" is a blanket term covering a broad range of four-wheeled interests, including hot rods, customs, imports and sports cars, street rods, and restored vintage cars, as well as oval-track racers, off roaders, modified trucks, and others. Enthusiasts come from all walks of life and every part of the country, and minor differences aside, their avocations are broadly similar. They read about cars. They work on cars. They attend shows, cruises, and races. They belong to car clubs, too—not broad-based groups like AAA, but clubs with sharper foci: sports cars, hot rods, racing, restoration. Collectively, they also spend billions every year on parts and accessories, because whatever their specific interests— foreign or domestic, on road or off, modified or factory-correct—theirs is not an economically rational approach to the automobile and its potential utility. Instead the car itself is paramount; *this* is what it means to be an enthusiast.

As we have seen, automobile enthusiasm has a long and varied history in the United States. For the purposes of this chapter, however, two specific developments require further attention. The first concerns the gradual embrace of salvage yards and out-of-service cars as automotive treasure, as well as the affinity many enthusiasts developed for junkyard scavenging itself, for the process of rooting through others' discoveries at swap meets, and for the sublime yet forlorn aesthetics of rows and rows of rusting relics in open salvage yards. The first trajectory of importance for the pages that follow, in other words, involves the fact that by the 1970s and 1980s, many gearheads had come to see wrecking yards and older-model cars in a manner decidedly out of sync with how most of their fellow Americans viewed them. The second trajectory is more straightforward: by the early 1970s, every enthusiast subgroup active in the United States had run into legal troubles of one sort or another. Drag racers and other motor-sports fans clashed with community groups and local officials who sought to curb noise pollution. Off roaders squared off with those who managed their public stomping grounds. High-performance enthusiasts— everyone from hot rodders and street rodders to vintage sports-car fans— regularly ran afoul of federal, state, and local emissions-control authorities. So too did restoration enthusiasts, who along with the owners of dune buggies and kit cars also encountered difficulties with titling, inspections, and vehicle registration.[35] In short, the enthusiasts of the 1970s increas-

ingly felt that they were under siege, for all around them the legal parameters of their automotive hobbies were rapidly contracting. Among
gearheads, this pessimistic and occasionally paranoid mind-set remains
pervasive to this day.

Here our two trajectories collide. For among the many legal challenges
enthusiasts have faced over the last forty to forty-five years, none have
been as fundamentally disruptive to them as a whole as have the broader
implications of the beautification effort launched in the 1960s. Many
enthusiasts struggled with emissions regulations, of course, and many
others encountered difficulties over their off-road trails, racing-venue locations, and other matters. But hobbyists of all stripes have had to grapple
with the antijunkyard, anti–parts car, and anti-old- or anti-rough-looking-
car sentiments that swept local regulatory systems in the 1980s, 1990s,
and 2000s. Put another way, precisely because they so highly valued
salvage yards, salvaged parts, parts cars, and older cars in general, and
precisely because they already felt as though the powers that be had singled them out for punishment, many gearheads were doubly despondent—
often, doubly furious—when their town halls clamped down on them in
the name of public beauty and property values. They were decidedly at
odds, that is, with the general-interest driving clubs that advocated roadside billboard and junkyard clearance.[36]

They seethed, for example, when new developments threatened long-
established salvage yards.[37] They also fumed whenever zoning changes
or new municipal bylaws forced enthusiasts with collections of parts
cars—both those with one or two parts cars in the suburbs and those
with full-blown de facto junkyards out in the country—to liquidate their
possessions.[38] Sometimes their anger stemmed from land-use changes that
happened behind closed doors and without notice to affected property
owners until the authorities arrived with summonses for newly minted
violations. As one Sullivan County, Tennessee, resident vented to Cars and
Parts in 2001, "I have 11 acres that was designated as a farm when I purchased it," and then "the county came through later and [re]zoned it as a
sub-division for tax reasons. I was never told of this until they had me in
court and was told to get all the cars off my property." Facing a penalty of
$500 per car per day, he had no choice but to move them. Later, he "talked
to everybody within ½-mile of me and they told me they also had not been
informed that their property zoning had been changed" and that he "was

not the only person whom they took to court and told to get their land cleaned up."[39] Similarly, others' frustration grew out of justifiable concerns over state-level proposals to allow municipalities to simply confiscate any vehicles they deemed to be junk *before* notifying their owners of any offense or planned action.[40] From time to time their anger also hinged upon the fact that the vehicles in question were not actually visible from the street or from neighboring properties; this was one of the key elements of the Mixon case.[41]

At other times, besieged enthusiasts were simply frustrated that they failed to persuade those in power that the parts cars and restoration projects in their crosshairs were not junk but rather part of the process of bringing a vintage vehicle back to life. This too was an essential aspect of the Mixon case, and it reflected a problem rooted in the earliest days of salvage-business zoning and regulation, the 1920s. Connecticut passed a landmark "Junk-Yard Bill" in 1929, for example, which was long before automobile restoration had developed into a hobby large enough to be noticed by the powers that be. Accordingly, the 1929 law assumed that any junk vehicles on any property must be there for the purposes of dismantling for profit and were regulated as such. Other states passed salvage-yard control laws in the 1930s and 1940s that were built around the same assumption. Fast-forward several decades, and the restoration hobby had emerged on a broad scale, along with the use of junked cars as restoration projects and parts cars. But the essential assumptions of these early junkyard-control laws remained unchanged: regulators in the 1980s and beyond still saw a "junkyard business" anywhere more than one weathered or wrecked car was parked.[42] Coupled with widespread municipal crackdowns associated with post-1960s beautification initiatives, these outmoded assumptions would remain a thorn in the enthusiast's side.

Finally, gearheads' anger, frustration, and sense of impending doom were often the result of the heavy-handed tactics the authorities used to combat perceived eyesores. Among the most egregious were the aforementioned ex post facto notice-of-confiscation and notice-of-zoning-realignment schemes. Others involved proposals to ban restoration work in home garages, to ban "ugly" trucks from residential areas, or, on a more widespread basis, to accelerate the retirement of older—and thus presumably ugly, out-of-date, unsafe, inefficient, or ecologically irresponsible—cars from the road through "clunker" or "scrappage" programs.[43] But those that stood out most of all were those in which the authorities themselves clearly bent the rules to try to force out a perceived eyesore. The Mixon case was one example, Andrew Zielinski's another. *Cars and Parts*, which long served (along with *Hot Rod* and a handful of others) as an unofficial regulatory watchdog for the enthusiast community, singled out two more in 1996.

The first involved North Baltimore County, Maryland, resident Michael Annen. When he moved to a two-acre property in a rural part of the county in the early 1990s, Annen brought his collection of seventy-four old cars with him. Some were restored, others rough; all of them ran. County officials were quick to cite the newcomer for failing to abide by regulations prohibiting the storage on private property of inoperable or unregistered vehicles. Since all of his vehicles were operable, Annen was able to obtain registration plates for them all, bringing his collection into compliance. Undeterred, the county commission proceeded to draft a new bylaw "limiting the number of cars parked outside on one's property to the number of bedrooms in the house on that property." Thus a two-bedroom home would be permitted two cars, a three-bedroom home three cars, and so forth. Annen of course did not possess a seventy-four bedroom house, leaving him and his collection in considerable legal jeopardy.

The second involved two enthusiasts living in a rural part of Knox County, Tennessee. One had a collection of thirty unrestored old cars, the other seventeen. County regulations required that all vehicles parked outside be in operable condition, and the local authorities were certain that one or the other of these two collectors were in violation—after all, their cars were old and arguably ugly. But when the county's inspectors came around to issue their citations, both gearheads in turn took a new battery around to each of their cars and fired them up. Flustered, the inspectors then convinced the county board to adopt a new requirement that every vehicle parked on private property be in operable condition and have its own battery. When the collectors met this new condition as well, the board passed further amendments to the regulations, requiring fully inflated tires, full registration documentation, and other stipulations. Regardless of their roadworthiness, large collections of old cars parked outdoors clearly were no longer welcome in Knox County.[44]

Small-time licensed salvage yards as well as genuine illegal junkyards rarely had much recourse when representatives from the county or township knocked on their doors. Larger salvage operations generally had a better track record of keeping the authorities in check,[45] but all too often since the 1960s they too have succumbed to urban renewal projects, rurban development, and other ventures. Individual enthusiasts, on the other hand, have long had an ace or two up their sleeves.

The first was the Specialty Equipment Market Association, or SEMA. Founded in 1963 as the Speed Equipment Manufacturers Association, by the 1980s SEMA represented those who manufactured, wholesaled, or retailed custom parts and accessories, everything from Jeep Wrangler roll bars and Honda Civic spoilers to high-performance camshafts, carburetors, and other equipment for cars of all types. SEMA also represented the enthusiast community itself, for without paying customers, the specialty-equipment industry would quickly have withered and died. Back in the 1960s and 1970s, SEMA saved the business and practice of hot rodding and other forms of high-performance motoring from regulatory oblivion when safety and emissions requirements first hit the books in the United States.[46] During the 1990s it worked as well to lead the defense of old cars, unsightly parts and project cars, and enthusiast collections of all kinds against excessive eyesore and zoning regulations. No less important, it also spearheaded the fight against the many federal, state, and local clunker or scrappage programs of the 1990s and 2000s.[47]

It did so with the aid of two subsidiaries, the SEMA Action Network (SAN), formed in 1997 as a grassroots watchdog and activist association, and the Automotive Restoration Market Organization (ARMO), formed in 1991, which represented specialty firms involved in the production of parts and systems for use in the restoration of older cars.[48] Through the SAN, SEMA was able to keep tabs on zoning, eyesore, and other measures coast to coast, enabling it to better utilize the skills of its D.C.-based attorneys and lobbyists.[49] Through ARMO, SEMA advertised to old-car hobbyists to warn them of the dangers lurking in their neighborhood associations, town halls, county boards, and statehouses. The SAN ran similar advertisements targeting broader swaths of the enthusiast community. One—"Don't Get Zoned Out!"—appeared in slightly different forms in *European Car*, *Eurotuner*, *Motor Trend*, *Hot Rod*, and *Popular Hot Rodding* between 2007 and 2009. The version that appeared in *Popular Hot Rodding* in April of 2007 is worth quoting at length:

You come home one afternoon only to find a ticket on your project vehicle that's parked on your property. Sounds like a nightmare scenario, doesn't it? But in some areas of the country, it's all too real. State and local laws—some on the books now, others pending—can or will dictate where you can work to restore or modify your project vehicle. Believe it or not, that cherished collectible you've hung onto since high school to pass down to your kids could very easily be towed right out of your yard depending on the zoning laws in your area. . . . To us, of course, these are valuable ongoing restoration projects. But to a non-enthusiast lawmaker, your diamond-in-the-rough looks like a junker ready for the salvage yard. If you're not careful, that's exactly where it will wind up.

It concludes by urging readers to join and actively participate in the SAN and thereby help SEMA keep their treasured projects safe.[50] SEMA also used its own periodical, *SEMA News*, to keep its member firms apprised of what was going on in cities, counties, and statehouses nationwide. Versions of these reports often made it into enthusiast periodicals as well, especially *Hot Rod* and *Cars and Parts*.[51] It's said you can't fight city hall, but SEMA, ARMO, and the SAN helped many do just that.

So did a second set of organizations. Some of these, including the World Organization of Automotive Hobbyists (WOAH) and the Council of Vehicle Associations (COVA), were founded in the 1990s as scrappage and zoning fears swept the enthusiast community. Both were loose federations of car clubs and other groups that were organized specifically to combat clunker programs, but they also assisted hobbyists with zoning and other matters. Other groups, including the Antique Automobile Club of America (AACA) and the Classic Car Club of America (CCCA), largely did the same.[52] Of greatest significance for the purposes of this chapter, however, was a group called CARZ: Citizens Against Repressive Zoning. During the 1990s CARZ worked with COVA and WOAH to sponsor seminars on land-use issues, town-hall politics, and other matters at car shows and swap meets. In the words of *Cars and Parts'* Bob Stevens, CARZ also "monitor[ed] individual disputes, disseminate[d] information on local zoning battles, compile[d] lists of attorneys active in such legal work, and generally serve[d] as a clearinghouse for information" on these and other matters. This gave many at the grassroots level the legal and organizational wisdom they needed to challenge the ex post facto NIMBYists and overzealous officials in their communities.[53]

Thus engaged, informed, and organized, enthusiasts did manage to ward off a number of local threats during the 1990s and 2000s. But because of its organizational breadth and its long experience with state and federal legislation, litigation, and lobbying, SEMA itself led the way in most of the broader, statewide battles of the 1990s and 2000s. In Ohio, it helped sink the confiscation-prior-to-notification proposals discussed above.[54] In nearly every other state—but most notably in West Virginia, Illinois, Maine, New Hampshire, and Oregon—SEMA also lobbied successfully for critical changes to de facto junkyard legislation to better protect enthusiasts with inoperable parts cars.[55] But its most celebrated victory came in 2005, when Kentucky governor Ernie Fletcher signed into law an inoperable-vehicles bill drafted by SEMA. Although it required parts cars to be shielded from public view of neighbors or street-level passersby, this new law secured an important guarantee: as long as they were shielded, inoperable cars could now be stored on private residential property throughout the state, regardless of the opinions of local authorities. Pundits often bemoan the passage of legislation drafted by lobbyists as anathema to American democracy, or as nothing more than yet another way to hand more to the wealthy at the expense of ordinary Americans. But Kentucky's landmark 2005 bill truly was a victory for the little guy: the Bluegrass State car enthusiast.[56]

None of this came easily, however, and few of the association's many victories were safe for long. When SEMA helped defeat a statewide de facto junkyard measure targeting enthusiasts in West Virginia in 2009, for example, it came after many years of legislative proposals, counterproposals, tabled measures, and zombie bills that returned from the dead with the start of each year's session. The same was true of many of the other ordinances and bills that SEMA, CARZ, and the enthusiast community challenged in the 1990s and 2000s. For in much the same way that gearheads have been loath to sacrifice their parts cars or their project cars, NIMBYists and beautification enthusiasts have been loath to give up their crusade against the "jalopies" and "junkers" they don't want to see in their backyards—or in their neighbors' backyards, either.

Wasting Well

In his final book, published posthumously as *Wasting Away*, the noted urban and environmental planner Kevin Lynch developed a conceptual framework for contemplating waste and wastefulness. Applied to the subject at hand, this framework suggests that our stories about junkyards, NIMBYists, and gearheads possess a much broader significance than their local details might imply. *Wasting Away* explores the phenomenon of waste in all its dimensions: wasted materials, wasted space, wasted time, wasted lives. Lynch argues that there are ways of *wasting poorly*—excessive consumption, inadequate waste-management systems, insufficient attention to toxic and radioactive materials—but also ways of *wasting well*, including managing the decline of Rust Belt cities, ensuring the inclusion of open space in sprawling suburbs, and paying adequate attention to properly sized and managed landfills and other means of disposal. Healthy cultures, he suggests, are those that effectively balance their *need* to produce certain kinds of waste with their *desire* to keep it out of sight and out of mind. In other words, generating too much waste is wasting poorly, but so is living in a world with too little.

To illustrate the point, Lynch deploys a pair of hypothetical dystopias. In one, continuous production and consumption in the name of a healthy economy mean that its inhabitants have stripped the planet of its resources (and those of the moon and several planets, too) and that their entire environment is supersaturated with trash. In the other, all forms of waste have been eliminated so that there is no trash and no air or noise pollution but also no free (wasted) time, no open (wasted) space, no (space-wasting) historical archives, and no (wasteful) unplanned births. Both of these societies are guilty, though in very different ways, of *wasting poorly*. On the other hand, at the level of the things we produce, consume, and discard, *wasting well* means living with a mix of things new and things old, things saved and things thrown away, things repaired and things replaced, things planned and unplanned.[57]

We have long lived in precisely such a world.[58] What matters for our purposes, however, is what we think of this world, and of its patterns of waste. For its part, the enthusiast community by and large embraces its mix of innovation, replacement, preservation, restoration, and adaptive reuse. Gearheads are not perfect, of course, not by a long shot. They burn fuel with impunity. They pollute. They are sometimes guilty of confusing public thoroughfares with high-speed raceways. But on balance, at least in Lynch's terms, they exhibit a healthy mix of the old and the new; they *waste well*. On the other hand, those behind junkyard-clearing urban renewal projects, those municipal inspectors who spend their days riding around in helicopters or verifying the roadworthiness of unrestored vehicles, and those who move next door to established salvage yards or enthusiast collections and then cry "Not in my backyard!"—these individuals offer a vision more in line with Lynch's second, overly sanitized dystopia. For they are guilty of propagating a fiction: that unblemished vistas coast to coast are not only possible, but also genuinely compatible with a consumer economy predicated, at least as far as the automobile is concerned, on Sloanist obsolescence.[59]

This remains a fiction on several counts, two of which are of immediate significance. First, it is a fiction because it pretends that millions of new Fords and Toyotas can materialize each year without the bones of their predecessors ending up *somewhere*, ruining *someone*'s vista. Second, and more to the point, it is a fiction because it pretends that those lovely antique, classic, and special-interest cars we all wave at and cheer for in Fourth of July parades and vintage-car races have somehow survived the years without requiring a junkyard here or a parts car there. Salvage-yard owners know better, and so do enthusiasts: somewhere, somehow, *someone*'s backyard must make room.

Then again, maybe not—not if we lived in a world without any older-model cars at all. As the final chapter of this book explores, more than a few policy makers in the 1990s and 2000s believed that this too was not only desirable, but possible.

6

Of Clunkers and Camaros

Policy Makers, Enthusiasts, and Old-Car

Scrappage, 1990–2009

Its panels were straight and clean. Its seats were worn but not worn out. Its motor still purred. It was a yellow 1967 Chevrolet Camaro, and in February 1994, *Hot Rod* unveiled it as its latest project car. Long a staple of enthusiast periodicals, project-car series normally unfold over several installments as the subject vehicle is modified or restored to meet a stated goal: period-correct cruising, weekend racing, reliable daily performance, or inexpensive horsepower gains.[1] But this one was different. Dubbed the "Crusher Camaro," its primary aim was political.

Hot Rod staffers rescued it from a queue of old cars in Los Angeles that Chevron planned to purchase and destroy to earn pollution-abatement credits from the authorities. The operation, "Project CAR,"[2] was an early example of what is variously known as a "clunker," "scrappage," "crusher," or, more formally, "accelerated vehicle retirement" program. Products of the federal Clean Air Act of 1990 and its attempt to ratchet up the fight against air pollution while easing the costs to smokestack industries and state officials, corporate- and state-backed scrappage programs sought to reduce pollution by permanently eliminating older cars from use, thereby accelerating the pace of vehicular obsolescence in the name of cleaner air. The Union Oil Company of California (Unocal) ran the first such program; Chevron and others followed its lead.

The concept was simple: pollution is pollution, whether mobile or stationary. If a firm has difficulty reducing its factory emissions, then why not allow it to tackle other regional sources that are more easily controlled? Or if a city or state cannot meet its federally mandated mobile-source pollution reduction goals, then why not allow it to think beyond automotive emissions testing and develop other strategies, including the deliberate destruction of older cars? This logic reflected a broader shift in the theory and practice of industrial ecology that was well under way by the late 1980s, a shift in the focus of pollution mitigation from the local (car tailpipes in Anaheim, factory smokestacks in Long Beach) to the regional or systemic (air quality in the Los Angeles basin as a whole).[3] The 1990 legislation embraced this shift by allowing industry and government officials to pursue systemic tactics, including old-car scrappage. So began the hunt for what politicians, corporate executives, and environmental regulators derisively labeled "clunkers" or "gross polluters." Precise definitions for these terms never materialized, apart from vague allusions to junky-looking, blue-smoke-trailing behemoths. But in practice, they referred to older cars of any stripe, condition notwithstanding—cars like *Hot Rod*'s Camaro.

That car was chosen for several reasons. Most simply, the circumstances surrounding its purchase meant that it was a vehicle with an interesting backstory, and thus an ideal candidate for a project series. More important, *Hot Rod*'s stated goal was to demonstrate that accelerated vehicle retirement was a methodologically flawed approach to air quality improvement because it assumed that *all* old cars, by definition, were clunking gross polluters. From beginning to end, the Crusher series aimed to refute this assumption empirically.[4] Most fundamentally, however, the Camaro's rescue was an act of defiance. By the end of 1993 a number of scrappage programs were in operation across the country, or were due to commence, and *Hot Rod*'s editors felt they had to take a stand. In doing so they joined a motley chorus of scrappage opponents, including scientists, skeptical journalists, and environmental and civil-rights activists. But *Hot Rod*'s opposition was different. By 1994, individuals, car clubs, and other organizations from across the gearhead spectrum had joined with automotive businesses and periodicals to argue that scrappage was not, as other opponents often claimed, a generally noble idea that was being executed poorly or that had too many unintended consequences. Instead, enthusiasts believed the deliberate destruction of older cars itself was wrong, both in theory and in practice. On the other hand, Unocal's annual report for 1990 included an essay *celebrating* old-car scrappage, complete with an image of a Pontiac Firebird—the Camaro's corporate General Motors twin—entering the maw of an automobile crusher.[5] For the Unocal brass, that

Firebird's status as a gross polluter was self-evident, and its removal from the streets desirable. But for enthusiasts its demise was tragic, and the liberation of its stablemate a righteous gesture.

Put another way, major stakeholders in the squabble over accelerated vehicle retirement were unable to frame the matter in common terms. For government officials, older cars were a loose end left untied when the environmental regulation of the automobile first began. For smokestack-industry executives, they were easy targets whose destruction cast their firms in an environmentally friendly light while enabling them to pursue a cheaper approach to industrial ecology.[6] For many green activists, this low-cost, corporate-backed approach was cynical and insufficient; the cars themselves were largely irrelevant. But for automobile enthusiasts they were precisely the point, and latter-day attempts to regulate them thus were seen as "tyranny," as Tom McCarthy notes.[7] This was because old cars of all sorts, everything from Cadillacs and Chevrolets to Triumphs and Volkswagens, constituted a "core aspect" of the individual and group identity of the enthusiast community. Organized plans targeting them therefore represented existential threats.[8] In other words, it was not simply a difference of opinion that kept those at Unocal, Chevron, or the Environmental Protection Agency (EPA) from grasping the value of a Firebird or Camaro or that prevented *Hot Rod*'s readers from comprehending the intent of scrappage advocates. Instead it was more fundamental: supporters and opponents of accelerated vehicle retirement simply were not speaking the same language.

This chapter details the emergence and interaction of these divergent views. It begins with a political and technological history of old-car scrappage, followed by an examination of the enthusiast's individual, collective, and institutional response. Along the way, it builds on a well-established body of historical and theoretical literature on obsolescence, waste, and recycling, as well as a robust policy-studies literature on accelerated vehicle retirement.[9] In particular, this chapter seeks to complicate our understanding of the meaning and power of obsolescence in the twentieth-century United States by juxtaposing its ultimate expression—the *enforced obsolescence* of accelerated retirement—with the countervailing worldview of the enthusiast. For although the automobile has long served as the poster child for an economic system predicated on a throwaway mentality and ever-shorter product cycles, its most ardent devotees have often seen things very differently. That is, where many see in old cars objects of transient value destined for the scrap heap, gearheads of all stripes tend to see classic cars[10] in the making—objects, in the words of social anthropologist and rubbish theorist Michael Thompson, of "durable value" that are worth protecting and preserving, not obliterating.[11]

Vehicle Emissions and Accelerated Retirement

During the 1950s and much of the 1960s, the federal government followed a wait-and-see approach to air pollution, and only with the Clean Air Act of 1970 did Washington assume a serious role in the fight against smog. Before then, California led the way. Its pollution-control act of 1960 set the stage for the first tailpipe standards, which debuted middecade. It also launched the nation's first air-pollution authority, the Motor Vehicle Pollution Control Board (MVPCB), a forerunner of the California Air Resources Board (CARB, established in 1967), itself a model for the federal EPA, set up in 1970.[12]

In its pioneering role, the MVPCB adopted a dual approach. First, it developed effective hydrocarbon (HC) and carbon-monoxide (CO) standards for model year 1966 California-market cars. This would be an enduring triumph for the board, its successor, and the state: following the Clean Air Act of 1970 and the arrival of federal EPA standards, California was for many years the only state allowed to set its own new-car rules, a nod to its groundbreaking 1960s work. Not everyone was pleased with these new agencies in Sacramento and D.C. Notably, automakers balked at their allegedly performance- and profit-sapping requirements, and by way of legal action, lobbying, and public threats of job losses, they managed to delay the onset of a number of new-car standards during the 1970s. But against a groundswell of bipartisan popular consensus that the measures were necessary, the automobile industry eventually conceded. By model year 1980, with the mobile-source provisions of the 1970 act in full force, showroom-new American-market cars ran much cleaner than ever before.[13]

The second part of the MVPCB's approach for California addressed older vehicles by requiring their owners to retrofit them with HC controls. CARB later followed suit, mandating oxides-of-nitrogen (NOx) devices for certain models. Neither move was popular among motorists, and during the early 1970s the state's authorities abandoned retrofitting altogether.[14]

Instead they settled on a simpler plan coupling ever-stricter new-car rules with periodic inspections of in-use vehicles to ensure they remained in compliance with their model years' standards: 1972 rules for 1972 cars, 1979 rules for 1979 cars, and so forth.[15] Because it grandfathered in the extant fleet, California's approach relied on the natural dynamics of vehicle turnover—that is, progressive obsolescence—to achieve improvements in the overall emissions of its millions of cars over time.

The federal, forty-nine-state approach was largely the same, and because Americans traded in their cars on a regular basis, it worked. With millions of new cars sold each year and millions of others retired to salvage yards, it was not long before the majority on the road were built since 1970, then 1975, then 1980. By 1985 the proportion of registered cars in the United States built before 1970 stood at only 6.7 percent, and by the early 1990s it had fallen to just over 2 percent. With a median age of 6.5 years, the cars of the late 1980s and early 1990s were overwhelmingly of the newer, pollution-controlled variety.[16]

Yet a growing number of policy analysts considered the remaining older cars a problem. Their statistics varied. Some claimed that vehicles built before 1971—prior to the advent of federal pollution-control standards—emitted thirty times more HC, CO, and NOx than newer models; a few pegged the figure as high as sixty-five. Others cast a wider net, blaming cars built before 1980 for more than 85 percent of air pollution in spite of their minority status within the national fleet. In time the claim that "10 percent of old cars make up 50 percent of the pollution" came to be the most frequently cited statistic, even though it was, in the words of a CARB official, an "urban legend." Nevertheless, the notion spread that older cars were hanging on too long and thus required an official nudge to oblivion.[17]

It was not a new idea. In February 1971, California governor Ronald Reagan told the Associated Press that "I have often wondered if we aren't going to come to a point where we are going to have to take a look at the possibility of funding and junking cars older than a certain age." The 50 percent figure was not new, either. One month after Reagan weighed in, Tom Carrell, the head of California's Senate Transportation Committee, agreed: "Getting old cars off the road is the only way to solve the problem. I'm sure that 50% of the auto smog problem is from the older car running on our highways."[18]

Nineteen years later their remarks were fast becoming policy, but not because of those who worried publicly about smog-belching cars. Instead, the impetus for putting accelerated retirement into practice came from those with smog-belching smokestacks—refineries in particular—who sought to sidestep federal, state, and local measures.[19]

Crushing Cars and Earning Credits

On April 26, 1990, Unocal announced that it planned to buy seven thousand vehicles in the Los Angeles area built prior to 1971 and permanently remove them from service by crushing them and selling them for scrap. In an advertisement the following day in the *Los Angeles Times*, the firm advised willing owners of qualifying vehicles that they would be paid $700 and given a one-month bus pass. Dubbed the South Coast Recycled Auto Program (SCRAP), the plan was hailed by many corporate executives, journalists, environmentalists, and politicians as a creative and forward-thinking approach to the air-pollution problem—a "win-win" for everyone involved.[20] Support poured in. T. J. Rodgers, CEO of Cypress Semiconductor, sent a letter of congratulations along with a check for $700 so that Unocal could "buy and bury one for us too." The California Community Foundation followed with a $70,000 donation, matched by Unocal. Spotting a retail sales opportunity, Ford contributed a $700,000 grant for the purchase of another thousand cars, and its Southern California dealers pitched in an additional $700 apiece to SCRAP participants who replaced their clunkers with brand-new Fords; First Interstate Bank reduced its new-car loan rates for program participants as well. Finally, the South Coast Air Quality Management District (SCAQMD), the pollution-control authority in the Los Angeles area, kicked in an extra $100,000.[21] In the end, Unocal was able to purchase and destroy nearly eighty-four hundred cars during the summer of 1990.

SCRAP was a speculative enterprise. Under fire for years from state and local officials for its refinery emissions, particularly those of its San Pedro installation near Long Beach, Unocal hoped that SCRAP would win the hearts and minds of the press and the public and thereby help convince the SCAQMD to develop and implement an emissions-credit system for the district. Were this to occur, the firm could turn its pilot program into a lasting effort, scrapping older cars (to Unocal, "gross polluters") for short-term stationary-source credits. This would buy the firm time to manage the more costly task of upgrading its aging refineries.[22]

Unocal's timing was perfect. While SCRAP was under way, Congress debated a series of amendments to federal pollution policy. Signed into law as the Clean Air Act of 1990, these changes recognized that urban air quality remained poor even after twenty years of effort. Marked improvements were evident in many areas, but during the 1980s scores of cities still failed to meet the 1970 act's benchmarks; this was especially true of Los Angeles, long the urban epicenter of unhealthy air. One important culprit was a disproportionate growth in annual mileage. Between 1970 and 1990, while its population grew by just over 20 percent, the annual number of miles driven in the United States nearly doubled, while in California it grew by 75 percent between 1980 and 1990 alone. This offset much of what had been gained through tighter tailpipe rules. Stationary polluters, although much improved since 1970, remained a major source of trouble as well, including the much-publicized phenomenon of acid rain.[23]

The Clean Air Act of 1990 attempted to address all of this by increasing the pressure on automakers, state and local officials, and industrial polluters. Notably, it required states with noncompliant metropolitan areas to submit to the EPA short- and long-term agendas called State Implementation Plans (SIPs), specifying how they aimed to bring their areas into compliance. The 1970 act had also relied on SIPs, but in 1990 their guidelines were modified to reflect a George H. W. Bush administration goal of incorporating greater flexibility into federal mandates: although SIPs were subject to EPA review, states were free to draft them in ways best fitting local conditions. For guidance, section 108 of the 1990 legislation listed a number of SIP options, including carpooling, transit improvements, and rideshare programs. Air-quality targets could also be met in part by implementing programs to earn SIP credits through "the voluntary removal from use and the marketplace of pre-1980 model year light duty vehicles and pre-1980 model light duty trucks."[24]

The inclusion of scrappage as an SIP option was largely the work of Senator William V. Roth Jr. (R). Earlier that year the veteran lawmaker from corporate-friendly Delaware introduced a bill (S. 2049) to allow automakers to earn credits toward their federal Corporate Average Fuel Economy requirements by paying for the destruction of older, presumably less efficient cars. S. 2049 went nowhere, and neither did a second attempt, S. 2237. But Roth continued to advocate scrappage as an ideal way to improve the fuel-efficiency and emissions performance of the national fleet. On the floor of the Senate he lauded the Unocal program, and with the help of Senators Max Baucus (D-MT) and John Chafee (R-RI), he worked to ensure that the compromise version of the 1990 Clean Air Act contained a scrappage option.[25]

Little noticed when the act was passed that November, the scrappage provision of section 108 first attracted attention the following spring, when President Bush directed the EPA and the Department of Transportation to develop an actionable SIP scrappage policy. William L. Schroer of the EPA's Energy Policy Branch took the lead and studied how a state-level program might work. His findings, circulated for comment in October 1991 and published the following March, supported the concept of vehicle scrappage as a means of reducing emissions in very large metropolitan areas, particularly those in warmer, drier climates like Southern California, where cars tend to stay on the road a bit longer than average. SIP scrappage assumed its final form when the EPA issued formal guidelines in February 1993, just after President Bill Clinton took office.[26] Thenceforward, states were officially permitted to establish accelerated vehicle retirement programs to help them meet their clean-air obligations.

President Bush had complicated matters the previous year, however. In early 1992 word leaked that the president, then a reelection candidate facing a tough primary challenge, planned to propose another, smokestack-industry-centered form of scrappage as part of a comprehensive energy strategy. His plan, dubbed "Cash for Clunkers," called for the establishment of systems for generating and trading mobile-source emissions-reduction credits (MERCs), which firms could use to offset their stationary-source pollution.[27] Bush's new plan prompted the EPA to study the MERC concept in addition to its SIP work.[28]

Preliminary guidelines emerged that spring, and pilot programs quickly followed. In 1992–1993, the U.S. Generating Company crushed 125 cars in Delaware, while in Chicago, seven oil companies teamed up with General Motors, the Environmental Defense Fund, and the Illinois EPA to destroy 207. In Philadelphia, the Sun Oil Company crushed 166 in 1993, while in Denver, Total Petroleum funded an operation in 1993–1994 in conjunction with a plan to subsidize the repair of salvageable vehicles.

Two hundred seventy-one cars entered Total's crusher; another 218 were repaired. In California, Unocal ran further iterations of SCRAP, and Chevron joined the fray with Project CAR.[29] Industrial demand for a scrappage option was strong, and in April 1994 the EPA finalized its MERC rules.[30]

A flurry of SIP and MERC proposals soon emerged within the forty-nine-state region, and "scrappage fever" ultimately surfaced in Arizona, Colorado, Florida, Illinois, Louisiana, Maine, Massachusetts, Michigan, New Jersey, North Carolina, Pennsylvania, Texas, Vermont, Virginia, and Washington—close to one-third of eligible states. But even with the benefit of considerable regulatory zeal, most of these states' plans died before a single car was destroyed. Pennsylvania erected the framework for a scrappage program in the early 1990s, but by the middle of the decade it had morphed into an emissions-repair subsidy. New Jersey passed a law requiring its Department of Environmental Protection to draft a hypothetical scrappage program, but Garden State legislators never moved to activate the plan. Regulators in Virginia began to toy with the idea of a scrappage program as early as 1992, and in 1996 their plans became law. But less than one year later, both houses of the Virginia legislature voted unanimously to repeal the program, and Governor George Allen agreed. Proposals in the 1990s in Colorado, Florida, Louisiana, Massachusetts, and Michigan never made it past the legislative process. Scrappage proponents tried several times to enact programs in Arizona and Vermont in the 1990s and 2000s, but they too failed, as did plans from the mid- to late 2000s in North Carolina and Washington.[31]

Of the many proposals floated in the forty-nine-state region in the 1990s and 2000s, only Maine, Illinois, and Texas got as far as crushing cars. Maine operated an ill-conceived and underfunded three-year program beginning in 2000, and the Illinois EPA launched a limited program based on its earlier pilot project, also in 2000. But only in Texas did a permanent program take root. Officials there developed an SIP in 1993, but the federal EPA rejected it in 1998. The following year, State Representative Warren Chisum (R-88th) tried three times to pass a MERC bill, but he too failed. State officials then devised a new plan more consistent with the EPA's rules, sidestepping the legislative process to deliver scrappage by bureaucratic fiat in 2000. Later modified to include an emissions-repair option, the Texas program—operated as part of the state's AirCheck emissions inspection system—is today the only active scrappage effort in the forty-nine-state region.[32]

Once hailed as an innovative policy option for reducing emissions, vehicle scrappage turned out to be at best a nonstarter and at worst an utter flop nearly everywhere it was tried. But before considering the reasons and forces behind its failure, we must briefly turn to California, the only place in the United States apart from Texas where accelerated vehicle retirement would remain a vital tool in the regulator's box.

California's Clunker Chaos

Unocal influenced federal policies with SCRAP, but the firm's primary aim in 1990 was to sway state and local officials in California. It would not be disappointed. In 1991 the SCAQMD revealed a new strategy granting stationary sources "more flexibility to reduce their emissions—and do it at the lowest possible cost."[33] Its main partner in exploring the idea was Unocal, and together they planned a SCRAP sequel for early 1993. SCRAP II targeted 1971–1979 models, half from 1971 to 1974 and half from 1975 to 1979. Thus its ostensible purpose was to compare the emissions of three categories of older cars: those built before 1971 (tested during SCRAP), plus the two groups chosen for SCRAP II.[34]

Unocal made arrangements with the recycling firm Hugo Neu-Proler for use of a car shredder, CARB for material aid with emissions tests, the office of the mayor of Los Angeles for a series of public-service announcements, and the SCAQMD, CARB, the Department of Motor Vehicles, and the Bureau of Automotive Repair (BAR, the agency responsible for California's emissions inspections) for financial assistance.[35] The SCAQMD also arranged for SCRAP II to serve as a dry run for a new rule reflecting its interest in providing flexibility to businesses in the fight against pollution. Rule 1610—publicly the "Regional Clean Air Incentives Market," or "RECLAIM"—allowed stationary-source industries to work with the district on programs, like SCRAP II, that emphasized alternative approaches to emissions reduction.[36] SCRAP II began in May 1993, and five hundred cars later, everyone had what they needed: CARB and SCAQMD officials their data, and Unocal $5 million in savings over three years in upgrades it otherwise would have had to make at one of its Los Angeles–area facilities.[37]

As far as the SCAQMD was concerned, the concept underlying Rule 1610 was proven. The vehicles tested and destroyed during SCRAP II were dirtier, on average, than brand-new cars by factors of 22 to 32 for HC, 15.5 to 18.5 for CO, and 4.9 to 6.5 for NOx. This was hardly surprising, since the newest car involved was a full fifteen model years old and was subject, when new, to much looser standards than those applicable in 1993. Compared with allowable levels for cars their age, those tested were 2.7 to 3.1 times dirtier for HC and 2.8 to 1.6 for CO, but cleaner by 30 to 50 percent for NOx.[38] Nevertheless, especially when viewed against the bleak data from SCRAP in 1990—one of the cars tested then was so dirty that its emissions alone could have powered a second vehicle[39]—these new data reinforced the notion that older cars were trouble, and industry-sponsored scrappage the solution. Accordingly, Chevron's Project CAR was quickly approved as a 1610 program in 1993, setting the stage for *Hot Rod*'s Crusher Camaro series. Unocal was given the green light to continue SCRAP as well, and by the middle of the decade the firm had established a subsidiary, Eco-Scrap, to manage its program and to dispose of vehicles for other Rule 1610 projects on a contract basis.[40]

Buoyed by the apparent success of Project CAR and SCRAP, the SCAQMD decided to expand the scrappage concept in the South Coast Basin, beginning with another mechanism, Rule 1623. Adopted in May 1996, this enabled area companies to earn emissions credits by scrapping gasoline-powered lawn mowers and compensating their owners with vouchers subsidizing the purchase of electric models. A popular 1623 program called "Mow Down Air Pollution" commenced in 2003, followed in 2006 by an effort to retire two-stroke leaf blowers in exchange for cleaner and quieter four-stroke models.[41] In 1995 and 1996 the SCAQMD also reworked an unpopular district ride-sharing mandate (Rule 1501) into a more politically palatable plan involving old-car scrappage. Rather than requiring local businesses to offer ride-sharing incentives to their commuting employees, the new system (Rule 2202) gave businesses the option of supporting a district-wide scrappage program instead.[42]

Agencies elsewhere in the Golden State developed programs of their own in the early to mid-1990s. The Ventura County Air Pollution Control District (APCD) operated a rideshare opt-out similar to the SCAQMD's, and privately financed plans were briefly run by the Sacramento Metropolitan and the Bay Area Air Quality Management Districts (AQMDs) as well. But most of the others did not involve emissions credits for private businesses and were instead public efforts solely intended to clean up the air. Programs of this sort run by the AQMDs or APCDs in San Diego, San Luis Obispo, the San Joaquin Valley, and Santa Barbara were up and running by middecade, and while some of these were short lived, an AQMD program in the Bay Area remained in operation through the end of 2010.[43] At the state level, the BAR developed a scrappage program tied to California's emissions-inspection system in the early 1990s. When it took effect later that decade, owners of qualifying cars that failed their emissions tests could opt to sell them to the state for disposal instead of paying for their repair.[44]

CARB and the state legislature also jumped in at the behest of the EPA. California authorities had long been free to develop their own emissions standards, but they were not allowed to let their state's air quality fall below the minimum standards applicable to other states. And in the early 1990s, even after several decades of effort, California's air quality still did not meet the federal standards of 1970, let alone those of 1990. CARB developed an SIP in 1994 to address this, calling, among other measures, for a statewide effort to scrap seventy-five thousand vehicles annually.[45] The following February, State Senator Charles Calderon (D-30th) introduced a bill, enacted later that year as S. 501, requiring CARB to run a one-year trial to test the feasibility of such a large-scale program. S. 501 also asked the board to draft new guidelines applicable to every scrappage effort in the state.[46] This was an attempt to bring some order to the chaos that prevailed by 1995, when multiple programs with different requirements and incentives were up and running from the Bay Area to San Diego. In particular, the Calderon bill aimed to harmonize the extant CARB approach—a set of guidelines it developed in the early 1990s guided, often loosely, the operation of AQMD and APCD programs[47]—with the BAR's statewide approach.

Following a series of drafts and public hearings, CARB released its preliminary regulations in December of 1998. After further hearings, the board then finalized them in the fall of 1999.[48] Or so it thought. A public, business, and interagency furor over CARB's new regulations erupted during 1999 and 2000, forcing the board to revisit its scrappage regulations in 2000 and 2001,[49] and again middecade.[50] Meanwhile, during 1998 and 1999, it also bought, tested, and crushed 1,001 1965–1990 model-year vehicles in a statewide pilot program designed by CARB and operated by a subcontractor.[51] Though CARB was satisfied with the results of this trial, funding for a pathbreakingly broad program of the sort envisioned in California's SIP would not materialize until the end of the 2000s, when the legislature called for the creation of a statewide "Enhanced Fleet Modernization Program" to target vehicles old enough to be exempt from the state's periodic emissions inspections. CARB was more than pleased to oblige.[52]

But by then, California's local and statewide efforts were anachronistic. Elsewhere, with the exception of Texas, accelerated vehicle retirement had long been on the wane.

The Rise and Fall of Old-Car Scrappage

For stationary polluters like Unocal, Total, and Chevron, accelerated retirement was all about saving money. Compared with upgrading their plants, buying and crushing a few thousand old cars here and there was an inexpensive way to reduce region-wide pollution, and they saw no reason why these reductions should not count toward their corporate emissions targets. The administrations of George H. W. Bush and Bill Clinton largely agreed, as did the EPA, the AQMDs and APCDs, and CARB.

But among these agencies, MERC scrappage was an acceptable approach less because it secured industrial cooperation and more because of a growing belief in the need to remove older cars from service. Indeed, the "old-car problem" had been a thorn in the regulator's side since the pollution-control approach centered on ever-stricter new-car standards became the norm in the early 1970s. That approach relied on a regular and relatively rapid turnover of the national fleet, and to a large extent it worked. The new cars sold by the end of the 1980s were far cleaner than those of 1970, and the national fleet was young enough by then that the air was demonstrably cleaner in many parts of the country, too—this in spite of the continued operation of (an ever-smaller number of) cars built prior to 1970. But the low-hanging fruit was also gone: all of the technologically simple and relatively inexpensive new-car gains had already been achieved by the time the debate over the Clean Air Act of 1990 commenced, and the EPA and the states found themselves chasing ever-more-fractional gains at ever-greater costs. In this context, it wasn't long before older cars resurfaced as a target.

This was especially true in California. Scrappage efforts there were often MERC-based, but a number of 1990s AQMD and APCD programs simply aimed to eliminate older cars. Definitions of "older" varied; some drew the line at 1981, others at 1971 or even 1959. But the underlying premise was constant: because they were not equipped with sophisticated smog controls, and because they had endured many years of wear and tear, they ought to be retired. The object was not to clean up older cars, nor to ease the burden for stationary polluters, but to accelerate the pace of obsolescence in the name of cleaner air.

Nowhere was this objective better articulated than it was in California during discussions regarding the fate of vehicles bought and crushed in 1990s scrappage operations. Should they be processed by the chain of actors who normally handle discarded cars? If so, that would entail parting them out to salvage useful components before their processing as scrap. Salvage-industry representatives had argued for this approach in other contexts,[53] but CARB instead demanded total vehicle destruction. Short of this, the board did not believe that SIP or MERC credits should be granted. When developing their statewide voluntary accelerated vehicle retirement guidelines in 1998 and 1999, for example, CARB officials argued that

> the only allowable use for any parts from a vehicle retired to generate emission reduction credits is as a source of scrap metal and other scrap material. The staff believes this approach best implements the basic principle of an effective [scrappage] enterprise—to accelerate fleet turnover to vehicles using cleaner, more advanced technology. Removing the *entire* vehicle from service not only eliminates the emissions of the retired vehicle, but it also renders that vehicle's parts unavailable for use in keeping another older, more polluting vehicle on the road longer than would normally occur through natural attrition.[54]

Here we have the most direct statement of the theory of accelerated vehicle retirement ever to appear in the regulatory corpus. For if the goal was to hasten fleet turnover to address the "old-car problem," then it made sense not to allow parts recovery. Policy analysts generally sided with CARB, though they sometimes differed on matters of detail and measures of cost-effectiveness.[55] The EPA's federal guidelines were somewhat looser, requiring the destruction of power-train parts but allowing the salvage of other components.[56] But for CARB, accelerated retirement was the primary goal, and it pursued this end with an exceptional focus.

The problem was that California was exceptional, too. With a fleet of twenty-three million, it was home to nearly 12 percent of all cars registered in the United States in 1990. Although this would drop to just over 10 percent by 2000, the Golden State still had far more cars than any other. It also had far more older cars, including 400,000 built before 1971.[57] This was because its low humidity and mostly salt-free winters inhibited rust, enabling its cars to remain in service much longer than those of Illinois or Massachusetts. Indeed, it was no coincidence that a state scrappage program nearly emerged in Arizona and did in Texas, two of a handful of states with desiccant climates similar to California's—and, in Texas, a sizable number of cars.[58] Elsewhere, natural attrition normally took care of the "old-car problem" before even the most ardent of scrappage advocates could reasonably call for action.[59] This was one of the reasons most state-level pilot programs failed to spawn sustained efforts.

Another was cost. During a trial run from 2000 to 2003, Maine offered those with cars built before 1998 new-car vouchers worth $2,000.[60] However, not only did it fail to provide sufficient funds for an effective level of participation—by mid-2001, only sixty-six vehicles had been destroyed—but it also failed to account for the costs of dismantling and disposal. Unlike many of California's scrappage programs, Maine's allowed for limited parts recovery, in accordance with EPA guidelines. According to salvage yard operators, however, most participating cars were too old to generate sufficient used-parts revenue to cover the costs of disposing their fluids and other hazardous materials, resulting in per-car losses of $300–$500. The legislature responded with an extra $350 per vehicle for participating yards, but now the state was on the hook for $2,350 per car, plus administrative costs.[61] As it floundered, Maine's experiment confirmed what a handful of policy analysts had long maintained: SIP scrappage made sense only in warm, dry states like California and Texas, places with very large numbers of old cars and emissions problems severe enough to render cost-effective the substantial public investment required.[62] Elsewhere, SIP programs made little sense.

On the other hand, MERC efforts were done in by a withering public assault. The concept had its advocates, including oil executives, Wall Street, President George H. W. Bush, and members of Congress from both parties. But the voices of dissent were more effective. Many environmentalists were appalled that crushing cars enabled corporations to avoid smokestack improvements,[63] while skeptical journalists questioned the economic and environmental logic behind old-car scrappage.[64] Also, in a direct challenge to the systemic approach to pollution mitigation, outraged civil-rights groups filed suit, condemning on environmental-justice grounds the notion that reducing mobile-source pollution in a given metropolitan area as a whole somehow made it acceptable for stationary sources—often sited in economically disadvantaged districts—to continue to pollute locally.[65] But by far the most vocal and effective critics of accelerated vehicle retirement were those who rallied around *Hot Rod*'s Camaro: automobile enthusiasts.

Crushing and the "Crusher"

In the 1960s and 1970s, gearheads of all stripes faced a number of novel regulatory challenges involving everything from beautification to noise-pollution mitigation, but those involving air pollution and highway safety were especially important. Regulations of this sort led high-performance enthusiasts in particular to rage publicly against the "ink-happy do-gooders" in government who threatened their cherished pastime by attempting to ban performance-enhancing modifications to newer, regulated cars. Behind the scenes, aftermarket industry organizations, especially SEMA, actually worked with these same "ink-happy" officials to arrive at compromises ensuring that in most cases, modifications to newer cars remained legal. But angry gearheads did not actively back these negotiations. Indeed, a 1970s attempt by SEMA to bolster its position vis-à-vis the authorities by establishing a subsidiary Enthusiast Division flopped for lack of gearhead interest. Nevertheless, by the end of the 1980s, most automotive hobbyists had calmed down. SEMA's tactics had worked, and their cars were safe.[66] Or so it seemed, until Unocal announced its bold agenda for the summer of 1990.

Angry letters from enthusiasts began to fly as soon as SCRAP began. The earliest went to Unocal itself, cut-up credit cards and boycott vows enclosed.[67] Later, especially after President Bush embraced the MERC concept, others voiced their rage in print. "Don't let your officials strip you of your car, your rights and your investments," warned one in the *Wichita Eagle* in 1992, while another sent an emotional defense of older cars to the *Buffalo News* the following year "on behalf of an old friend, my 1967 Ford convertible." A more substantive argument appeared in the *Chicago Tribune* in 1992, when the president of a restoration group wrote that even if there were an "old-car problem," with a little patience natural attrition would resolve it without the expenditure of taxpayer dollars or the allowance of stationary-source credits. A less robust letter in the *Los Angeles Times* prompted Unocal's boss to respond; later, several enthusiasts reacted angrily to a pro-MERC editorial in the *Wall Street Journal*.[68] Scrappage skeptics on mainstream newspaper staffs soon joined in with contributions questioning accelerated vehicle retirement on economic and environmental grounds.[69] Others reported on the automotive hobby's reaction to vehicle scrappage plans, correctly noting in one case that "to auto collectors, destroying an old car is like burning a Rembrandt."[70]

Automotive periodicals were also abuzz. Letters to the editor called on gearheads to unite in defense of older cars and to boycott businesses supporting scrappage. Editors lambasted the MERC concept, warning that what is voluntary today could easily become mandatory tomorrow and criticizing the assumption that all old cars are dirty.[71] *Car Craft* was the first to weigh in editorially, in 1990, followed in 1991 by *Hot Rod*, long the world's best-selling automotive periodical. Specialty titles like *Hot VWs* and *Chevy High Performance* soon joined in, and by the mid-1990s even broader magazines like *Motor Trend* and *Car and Driver* had voiced concern. Curiously, the restoration-oriented *Cars and Parts*, whose readers stood to lose the most as scrappage fever spread, said virtually nothing about the subject until 1993; thereafter, its depth of coverage rivaled even that of the industry-insider *SEMA News*.

Unocal was taken aback by all this gearhead outrage. "They just won't listen to us, and we've told them over and over that we're not crushing classics," said a company spokesperson in 1994. "There are no Ferraris in this program." Senator Roth was equally stunned by the angry letters he received after praising SCRAP in 1990. To him, accelerated retirement was all about clunkers, not collectibles.[72] What Unocal and its federal champion failed to realize was that even the least desirable sedans of the 1960s and 1970s were nonetheless of value to 1990s enthusiasts, especially hot rodders and restorers, as sources of major drive-train and other components. Unocal's straight-to-the-crusher approach, as well as CARB and EPA rules prohibiting power-train-parts recovery, therefore threatened shiny classic cars with engine maladies and rusty, unloved family cars alike.[73]

More to the point, "classic" and "collectible" were designations constantly in flux, no less for old cars than vintage furniture, paintings, and books. As Michael Thompson explains in *Rubbish Theory*, shifts over time in the perceived value of material objects are common and tend to occur in graduated stages. When an item is new it resides in the first, *transience*, where its symbolic and monetary value diminish with time. Eventually it enters the second, a virtually worthless state of limbo known as *rubbish*. Then the object either disappears entirely or is "discovered" by a creative elite of eccentrics who facilitate its passage into the third stage, *durability*. Durable objects are those of lasting value or historical significance: art, antiques, decorative tapestries, and the like. Although he admits that older cars occasionally become collectible, for the most part Thompson identifies the automobile as the quintessential transient object destined for oblivion.[74]

Here he is mistaken. As Thompson himself explains, those who are around when an item is transient tend to have difficulty imagining it as an object of durable value, whereas those who enter the picture later, after the item in question has already become rubbish, often have no trouble recognizing its collectability.[75] Small wonder, then, that Thompson failed to anticipate that the secondhand cars of his time, the 1970s, would ever be desirable. Automotive history is rife with similar moments. Just after World War II, enthusiasts actively debated the collectability of 1930s Cords and Packards, but ten years later their desirability was no longer questioned. Likewise, in the early 1970s it was not yet clear whether the domestic cars of the mid-1950s would ever be cherished, but by the end of the decade they certainly were. Similar shifts occurred in the late 1970s and 1980s regarding 1960s muscle cars and other vehicles, and by the early 2010s some within the hobby had even begun to embrace the sedans of the 1970s in their own right.[76] In short, cars are no different from armoires or coins, for among them, "collectible" or "classic" never can be firmly fixed for all time.

So when a lawmaker from Washington State unveiled a proposal to scrap old cars in 2006, he missed the point when he defended his plan by claiming that he was "not going after the classic '57 Chevy," but instead "the Plymouth Duster." Even if Dusters had not yet come into vogue as inexpensive starter projects—and by 2006, they had—the underlying premise still was faulty.[77] Cars of little interest to today's enthusiasts could be of substantial value to tomorrow's, and this was a critical reason why so many disliked scrappage. For had it surfaced in the 1960s, it would have obliterated 1957 Chevys—or, if in the late 1940s, 1937 Cords. To the enthusiast, it was no more acceptable that 1970s Darts and Vegas were in the clunker crosshairs in the 1990s.

Not every gearhead was outraged when confronted with accelerated retirement proposals, however. Some urged their fellow travelers to breathe easy, since the low bounties normally offered meant that few if any classics were in jeopardy. After all, what enthusiast would part with a beloved car for a measly $700? Instead, those crushed would be genuine clunkers, typically 1970s behemoths worn out after years of abuse and ready to be put to sleep.[78] But this was a minority opinion. To most the risks were real, in part because of the aforementioned uncertainty regarding the identity of tomorrow's classics, but also because many examples of the cars already cherished by hobbyists in the 1990s were at the time still owned by motorists who simply viewed them as expendable transportation—cars that had not yet fully escaped from the low-value limbo of Thompson's rubbish phase. Unocal's spokesperson did little to address the resulting fear among gearheads, declaring in 1990 that "as far as discriminating between a collector car or a regular old car, we will pay the $700 for each one, and they will be scrapped."[79] Enthusiasts responded with disbelief and outrage.

In fact, their heated rhetoric often echoed that of the late 1960s and early 1970s, when safety and emissions standards first posed a threat to hobbyists, especially those who modified their cars. As we have seen, aftermarket industry organizations successfully met that challenge, saving a large part of the enthusiast community in the process. But of equal importance was what those who fought this early battle learned about the regulatory process, and how to influence it. SEMA emerged as a powerful lobbying force, and enthusiasts were generally more aware that it wasn't enough to simply oppose an idea harmful to their pastime. One also needed to come up with a more palatable alternative—to work *with* regulators and legislators, that is, to arrive at mutually agreeable compromises.[80] Accordingly, the *action* of the early 1990s often echoed that of the 1960s and 1970s as well.

Consider the Crusher Camaro. Ostensibly a magazine project car with an unusual backstory, it was also a politically motivated project designed and executed in a manner reminiscent of the most important contributions that prohobby periodicals made during the emissions and safety squabbles of decades past. By nature a hands-on lot, car magazine editors back in the early 1970s frequently made their case that modified cars were not necessarily dirty or unsafe by building demonstration models for the benefit of their readers and the regulatory community. In 1971, for example, the editor of *Hot Rod Industry News*, Don Prieto, built a high-performance small-block V-8 that ran cleaner when modified than when in stock condition. Shortly thereafter, *Hot Rod*'s Bob Weggeland and the Offenhauser Equipment Company did the same, assembling a powerful and cleaner-running V-8 for a 1968 Corvette. *Popular Hot Rodding*'s staffers then built a high-performance engine for a 1967 Chevrolet that was capable of passing the emissions test for model year 1973 cars. And in 1974–1976, *Hot Rod* worked with SEMA on a large-scale project called the Combination Testing Program, in which a number of vehicles with a variety of engines underwent a series of hot-rod modifications and almost always wound up cleaner-running. The Combination Testing Program in particular served as an effective tool in making the hobby's case to the EPA.[81]

Similarly, the Crusher Camaro project aimed to demonstrate that the old cars targeted by Unocal, Chevron, and others were not necessarily gross polluters. *Hot Rod*'s David Freiburger and Rob Kinnan spearheaded the project. After buying the car from an individual who planned to sell it to Chevron's Project CAR, they headed first to a state-approved emissions-testing station—one that was run by Chevron, no less. There the car easily earned a passing score. Then they replaced its six-cylinder engine with a V-8 before installing a number of performance-enhancing parts. Further testing upon completion showed that it was clean enough to pass the tailpipe test for 1979 model-year California vehicles—not bad for a car produced in 1967!

Nevertheless, the Crusher actually failed its emissions test, for in California, an under-hood inspection has long been a routine part of the process. Few of the modifications the *Hot Rod* team performed were legal by the letter of the law because the parts selected did not carry CARB's official seal of approval. In other words, as far as the State of California was concerned, it did not matter whether a given vehicle's tailpipe emissions were as clean, or even cleaner, than required. For CARB and the BAR, tested vehicles had to run clean *and* do so with stock or otherwise approved parts. Freiburger and Kinnan went on to use the case of the Crusher's emissions results to lobby against the California under-hood requirement, although they failed to convince the authorities in Sacramento.

In terms of the scrappage debate, however, the Crusher was indeed "a success on all points." It was not a clunker. It was not a gross polluter. Yet were it not for *Hot Rod*'s intervention, Chevron would have obliterated it in the name of cleaner air—and would have won the right to pollute a little more itself in the process.[82]

Fighting Scrappage Fever

The Crusher showed the way, but activist gearheads also advanced broader points. Some questioned the rigor with which their opponents factored annual mileage and the economic status of program participants: might it not be true that many of the vehicles targeted by scrappage advocates were seldom-driven second or third cars belonging to solidly middle-class people, rather than genuinely out-of-tune clunkers driven every day by those who could afford neither to maintain nor to sacrifice them? Echoing skeptical journalists, enthusiasts also claimed that pulling serviceable examples out of circulation inflated used-car prices, disproportionately harming those of lesser means. Others cast doubt on the computer models used to predict scrappage-related emissions gains, questioned the wisdom of granting MERCs to businesses, and tirelessly emphasized that today's clunkers could well be tomorrow's classics.[83]

But their most significant arguments were three. The first centered on *manufacturing equity*: once the environmental costs of building and shipping new cars are included, any net emissions gains from scrapping older models can take more than a decade to materialize.[84] The second focused on programs that did not allow parts recovery. Here the claim was that this approach did not actually hasten the retirement of similar cars, as CARB contended. Instead it was more likely to lead to an *increase* in pollution, as vehicles that could have been repaired with scrapped parts instead were driven without being fixed at all.[85]

Finally, and most fundamentally, they railed against the notion that old cars are by definition dirty. One example often cited to support this claim involved an early 1990s Southern California roadside spot test. Three cars were stopped and tested, a recently serviced 1985 Honda Accord, a 1985 Ford Bronco in superb form, and a 1974 Cadillac Coupe de Ville with torn upholstery, missing chrome, and layers of grime. The cleanest of the three? The Cadillac, by far. The others, though newer and certainly shinier, had emissions profiles far outside the allowable parameters.[86] The moral was simple: regulatory activity should focus on actual gross polluters—defined as cars with genuine emissions problems—regardless of vintage or outward appearance. The EPA agreed, declaring in 1992 that "old cars are not always, or even uniformly, high emitters. Rather, their high average emissions generally are due to a small set of extremely high emitters, while other old cars are only moderate emitters, and some are quite clean."[87] But this was a tough sell against the entrenched public image of the old, smog-belching clunker.[88] To make the argument that vehicle age and apparent condition simply did not matter, the hobby needed a stronger, collective voice.

Local, regional, and national organizations shone in this regard. Long-established, large-scale car clubs, as well as regional and national umbrella organizations, focused the grassroots energy of millions of ordinary enthusiasts on effective lobbying. In Pennsylvania, enthusiasts rallied in Harrisburg to try to convince state lawmakers to abandon the state's plans for a scrappage program in favor of an emissions repair subsidy. They succeeded.[89] Similarly, the Southern California Galaxies Club worked with *Hot Rod* in its early efforts to get the word out about SCRAP, while the Association of California Car Clubs, the Association of Car Enthusiasts (upstate New York), the Arizona Automobile Hobbyist Council, the New Mexico Council of Cars, and the Texas Vehicle Club Council drew up petitions and lobbied their respective statehouses.[90] The Antique Automobile Club of America, the Council of Vehicle Associations, the World Organization of Automotive Hobbyists, the Antique Auto Coalition, and other nationally oriented groups—some old, some new—held seminars on regulatory activity and political activism at car shows; some of them even met with federal officials, both to gauge the mood of the new Clinton administration and to make the concerns of the enthusiast community known to those at the top.[91] Businesses with a lot to lose if old-car scrappage became routine also stepped up. Len Athanasiades of Year One, a restoration parts supplier, distributed pamphlets on antiscrappage activism, while ardent enthusiast (and environmental activist) Terry Ehrich, the publisher of *Hemmings Motor News*, offered matching funds to those "who create programs or engage in efforts to protect, preserve and promote the collector-car hobby."[92] Other magazines got involved as well, notably California-based *Hot Rod* and Ohio-based *Cars and Parts*.

Industrial organizations spoke out, too. Some joined the antiscrappage side of the debate because they represented businesses involved in the recovery, rebuilding, or resale of used parts—businesses that were concerned about the stance of certain regulatory bodies on the matter of parts recovery from scrapped vehicles. This was true of the Automotive Parts Rebuilders Association, the Automotive Engine Rebuilders Association, and the Production Engine Remanufacturers Association. Others, including the Automotive Warehouse Distributors Association, the Automotive Service Industry Association, the Motor and Equipment Manufacturers Association, and the Coalition for Auto Repair Equality, joined the fight because their constituents repaired or supplied parts to repair vehicles and were therefore concerned about the impact of programs that encouraged scrapping cars rather than fixing them. Vocal as well were bodies representing those who catered chiefly to enthusiasts by supplying components to restore or modify cars of all ages. Here the most active were the Automotive Aftermarket Industry Association, the Street Rod Equipment Alliance, the Automotive Restoration Market Organization (ARMO, a SEMA subsidiary), and SEMA itself.

From time to time these organizations lobbied jointly,[93] but in the long run SEMA played the most visible and effective role. This is not altogether surprising given the association's long record, dating back to the mid-1960s, of successful action in the regulatory realm. As a result of this history, SEMA officials had extensive federal, state, and local contacts and were better equipped than most to participate in this latest legislative battle. Accordingly, its California headquarters as well as its legal counsel in Washington, D.C.—and later, a satellite D.C. office—served as information clearinghouses. SEMA also sponsored studies to probe the assumptions of scrappage proponents; submitted detailed technical and legal comments on clunker schemes; attended public hearings at the EPA, CARB, and other agencies; and lobbied state and federal lawmakers directly. SEMA and ARMO ran advertisements warning enthusiasts of the dangers posed by scrappage programs, and SEMA developed an overarching media strategy to guide those in the publications business who sought to reach out to enthusiasts.[94] SEMA also worked to cultivate a direct relationship with hobbyists. This it accomplished in 1997 by establishing the SAN. A bottom-up web of enthusiasts, businesses, and clubs, the SAN served as both a means through which SEMA's Washington office could rally the enthusiast community at critical moments and as a way for SEMA officials to better monitor local developments.[95]

Of the studies SEMA sponsored, two were of particular significance. The first was conducted by its director of technical affairs, Frank Bohanan, in 1992. Bohanan's objective was to confirm empirically what most enthusiasts knew in their guts: old cars are not necessarily gross polluters. Using recent emissions-testing data on more than one million vehicles from California's BAR, Bohanan found that "*the dirtiest 10–20 percent of 1980–1984 model year cars actually emit higher levels of HC and CO than the majority (cleanest 50–60 percent) of most older vehicles.*" He then collated these findings with odometer readings from the Department of Motor Vehicles and determined that the oldest cars in the sample were also driven the least, further reducing their contribution to pollution in the area. In addition, CO_2 levels (a reliable measure of fuel economy because CO_2 is directly proportional to fuel use) were worst among the *newest* cars, the 1992 models. But his most striking discovery was that the worst of the worst were not among those built before 1971, nor 1981. Instead, genuine gross polluters overwhelmingly belonged to a newer range of models typically exempt from scrappage plans: 1980 to 1984. SEMA made much of these data in its written and oral testimony.[96]

The second was a joint study by SEMA and the San Diego APCD of the efficacy of subsidizing the repair and upgrading of gross-polluting cars rather than their destruction. Between 1996 and 1999, the authorities in San Diego offered vouchers to residents whose cars failed their smog tests, subsidizing their repair and retrofitting them with inexpensive pollution-control devices. Tailpipe emissions were reduced by an average of 60 percent, more than enough in most cases for the cars to pass their smog tests and run cleaner than other vehicles their age. Because it did so at a total cost of approximately $500 apiece, as opposed to roughly $1,000 per vehicle scrapped, SEMA hailed the pilot program as a cost-effective success and lobbied vigorously in the years that followed for "repair and upgrade" as an alternative to accelerated retirement.[97]

Using these and other data to bolster its position, SEMA used every institutional and legal tool at its disposal to combat the spread of scrappage programs. Among several other fronts, the association worked with *Hemmings*, Year One, Petersen Publishing (*Hot Rod*'s publisher), several California car clubs, and CARB itself to draft a list of collector vehicles and engines in 1993, which CARB then recommended for exemption from AQMD and APCD scrappage programs. Not everyone within the enthusiast community was pleased with this list—some were concerned that it set in stone the traditionally fluid definition of "collectible"—but it was a step in the right direction in terms of regulatory cooperation.[98] SEMA also lobbied, in California and elsewhere, for waiting periods for vehicles in the scrappage queue as well as phone-in (later, e-mail-based) notification systems to give interested parties a chance to save selected cars from the crusher. This effort succeeded first with the Illinois EPA, during the period when it actively contemplated a permanent scrappage program, and later with CARB.[99]

SEMA also lobbied hard against the establishment of a scrappage program in Texas in 1999, helping defeat State Representative Warren Chisum's legislation three times that year. But this was a short-lived victory: the following year, as mentioned above, the state's pollution control authorities bypassed the legislature and implemented an accelerated vehicle retirement program on their own. SEMA's successes in California were also tempered by occasional defeat. When CARB developed its statewide scrappage guidelines in the late 1990s and early 2000s, it ignored the concerns of SEMA, the automotive salvage industry, and, indeed, the letter of California law (S. 501) by prohibiting parts recovery from scrapped vehicles. Despite considerable pressure from state lawmakers and industry associations, CARB stood by its ideologically pure stance on accelerated vehicle retirement.[100] Elsewhere SEMA enjoyed more lasting success. In Arizona, its Action Network helped defeat a MERC plan in 1997 in spite of the deep pockets of its chief supporter, the Western States Petroleum Association. The following year, Arizona implemented a SEMA-backed "repair and upgrade" program instead.[101] The association's efforts also helped quash accelerated vehicle retirement proposals in Vermont, Colorado, North Carolina, and several other states in the 1990s and 2000s.[102]

SEMA also beat back numerous attempts to establish federal subsidies for state-level programs. Most of these efforts centered on Congestion Mitigation and Air Quality (CMAQ) funds, a product of federal legislation passed in the wake of the Clean Air Act of 1990. Federal CMAQ funds could be used for a variety of state programs to ease traffic and pollution, including transit and carpooling, but not accelerated retirement. Scrappage advocates tried to lift this restriction on nearly a dozen occasions in the 1990s and 2000s, but SEMA, the Council of Vehicle Associations, the Automotive Service Association, and other groups prevented this from happening each and every time.[103]

The fact that lawmakers kept trying to repeal the CMAQ ban might appear to suggest that scrappage was alive and kicking well into the 2000s. And in some respects, it was. But in the documentary record, the recurring debate over these funds appears increasingly out of sync with the times. For by the early 2000s, most of the states that once toyed with scrappage had rejected the concept in favor of cheaper, less politically volatile approaches. What's more, by then the natural pace of obsolescence had driven even more of the cars of the 1960s, 1970s, and 1980s from the roads, making it less and less possible to argue credibly that destroying them remained a viable approach to the problems highlighted by the Clean Air Act of 1990. SEMA was by no means alone in helping legislators and regulatory agencies arrive at this conclusion. Other industry associations, umbrella car-club organizations, local clubs, ordinary enthusiasts, and monthly periodicals played vital roles as well. Their defense of the old against the tide of the new may not have convinced everyone, but it did hold America's scrappage fever at bay, long enough in most cases for it to lose the seemingly unstoppable momentum it had in the early 1990s.

Nevertheless, as a policy tool, accelerated retirement refused to exit quietly.

Clunker Nation

Mired in the "Great Recession" in early 2009, federal lawmakers began to discuss additional means of stimulating the economy, particularly the automobile industry. General Motors and Chrysler were teetering on the brink in spite of massive infusions of taxpayer dollars, while Ford had mortgaged everything—including its trademark oval—to keep itself afloat. Inventories of unsold cars and trucks piled up at dealerships, and production slowed to a crawl.

Inspired by recent European experience using accelerated retirement to jump-start new-car sales, competing stimulus proposals rooted in accelerated vehicle retirement schemes emerged on Capitol Hill in the spring and early summer. Representatives Betty Sutton (D-OH), Candice Miller (R-MI), and Bruce Braley (D-IA) introduced a stimulus-scrappage bill in the House, and competing plans from Bob Casey (D-PA), Debbie Stabenow

(D-MI), and Sam Brownback (R-KS) and a group spearheaded by Dianne Feinstein (D-CA) were introduced in the Senate. During the summer of 2009, these bills were hotly debated in the halls of Congress and in the court of public opinion. Newspapers published opinions for and against the idea, enthusiast periodicals once again railed against the dreaded crusher, and SEMA bumped its efforts into high gear.[104] In the end, President Barack Obama endorsed a hastily drafted compromise, the Consumer Assistance to Recycle and Save Act (CARS, popularly known as "Cash for Clunkers").

Operated as the Car Allowance Rebate System in mid- to late 2009, the program offered generous vouchers to those willing to replace their "clunkers" with brand-new and more fuel-efficient cars. Parts recovery was allowed, except for engines, and in a radical departure from 1990s norms, the program targeted *post*-1984 models. This was SEMA's doing. Well aware that public opinion overwhelmingly favored the development of some sort of scrappage program as a form of automotive-industry stimulus, SEMA focused its lobbying efforts in early to mid-2009 less on stopping the inevitable than on making sure that whatever eventually emerged from the 111th Congress would at the very least exclude collector vehicles and allow parts recovery.[105]

In the immediate short term, CARS was a success, as eager participants — "clunker nation," they were sometimes called—moved thousands of cars off dealer lots and removed thousands of others from use. In particular, the program managed to eliminate a number of the heavy and inefficient SUVs sold in the halcyon days of the late 1990s and early 2000s, when fuel prices reached historic lows. But the apparent achievements of the CARS program were little more than an illusion. Immediately after it ended, and especially in early 2010, new-car sales slumped again, and to an even deeper level than before. This was because consumers who had planned to buy new cars in late 2009 or early 2010 instead had simply shifted their purchases up by a few months in order to take advantage of the government's largesse.[106]

But the ultimate success or failure of the program mattered less than its form. CARS had nothing to do with air quality, nor with older cars. Instead it aimed to aid an industry in crisis by directly subsidizing new-car sales, and that it hinged on fuel efficiency rather than emissions or vehicle age confirmed the demise of accelerated retirement of the 1990s variety, at least at the federal level. More precisely, it confirmed that scrappage had been transformed into a tool for economic stimulus: accelerated obsolescence for the good of the economy rather than the good of the air.

Clunker nation therefore brings to mind the work of the midcentury social critic Vance Packard. *The Waste Makers*, his third best seller, begins with a dystopian vision of the city of the future, a place where nothing is made to last. Instead, every aspect of life in "Cornucopia City" is geared toward the continuous disposal of the old in order to allow for the uninterrupted and profitable consumption of the new. Thus, after four thousand miles or one year's service, cars there are considered "old," and in one particularly revealing passage Packard describes how motorists who surrender these cars for disposal are "rewarded with . . . one-hundred-dollar United States Prosperity-Through-Growth Bond[s]."[107] Squint a bit and Packard's text could well describe 2009, when close to 700,000 cars were scrapped before their time to keep the Motor City running.

For enthusiasts, of course, "before their time" is largely beside the point. In their view serviceable, restorable, or otherwise useful cars of any vintage ought never enter the maw of the crusher as a matter of public policy. Tellingly, however, few gearheads protested with much conviction the deliberate destruction of late-model Jeep Grand Cherokees and Ford Explorers in 2009. Following an initial burst of outrage when the idea of a cash-for-clunkers stimulus first surfaced, their anger largely abated once it became clear that CARS excluded anything built before 1984, no matter how decrepit.[108] Perhaps, as Michael Thompson might suggest, 2009's enthusiasts were unable to perceive those recent-vintage Jeeps and Fords as anything other than transient objects destined to become rubbish. Perhaps, that is, they were unable to recognize in the cars of the 1990s what many of them clearly saw in those of the 1970s: future classics.

Perhaps. On the other hand, despite their theoretical objections to the euthanization of cars of any age, many enthusiast subgroups had long since given up on new models anyway. This was more than a temporary "buyers' strike" of the sort seen briefly among mainstream consumers at the end of the 1950s and instead marked a more lasting rejection of the showroom-new.[109] It was true not only of avowedly preservationist old-car restorers, but also many hot rodders, sports-car fans, and others. Some of these contrarians gave up on new cars during the early to mid-1970s, when automakers rolled out underwhelming American-market cars saddled with makeshift emissions controls and heavy, energy-absorbing bumpers. Others gave up later, when computerized ignitions and fuel-injection systems took the fun out of shade-tree tuning. Some among the disenchanted blamed the automobile industry for no longer building desirable cars, while others blamed the National Highway Traffic Safety Administration, the EPA, and CARB for forcing firms like Ford and General Motors to give up on power, speed, and style. Either way, the new-car showroom had ceased to be an unambiguous draw for each and every red-blooded American gearhead.

As a result, during the 1970s, as it fought to preserve the practice of new-car engine and chassis modification, SEMA failed in its attempt to recruit the aid of enthusiasts. During the 1990s, however, when old-car scrappage was the pressing issue of the day, enthusiasts joined its Action Network in droves. Their activism then receded somewhat in 2009, when CARS went after newer models. This muted response to clunker nation contrasts sharply with the gearhead community's longer history of spirited resistance to accelerated vehicle retirement. Indeed, it suggests that a sizable number might well have agreed with what one turn-of-the-century hot rodder saw as the dominant attitude of the time: "They stopped making cars in 1972, dude."[110]

Coda

Something Old, Something New

They stopped making cars in 1972, dude. Why 1972? The outlook that our fin de siècle hot rodder described was fairly straightforward: those built after model year 1971 were strangled by government mandates, many believed, and thus were less quick, less fun, and less attractive. Accordingly, those of an earlier vintage were more desirable—even collectible. But age alone was only one of many factors governing desirability and collectability. Scarcity mattered as well, as did market values. But sometimes neither scarcity, value, nor age entered into it at all. Ford Mustangs and Volkswagen Beetles were affordable when new, and both sold in the millions. Both became collectable by the middle of the 1970s, too—before the end of U.S. Beetle sales and just a few short years after the first-generation Mustang was discontinued.[1] In spite of their relatively recent vintage, their abundance, and their reasonable market values, both spawned vibrant enthusiast subgroups that persist to the present; both have inspired more than their fair share of treasure hunts and barn finds over the years as well.[2] The same was true of Camaros, MGBs, and a number of other less-than-ancient, less-than-scarce, and less-than-budget-breaking collectibles. Clearly, there must have been something else that set these cars apart.

Indeed there was, but it wasn't necessarily anything in or of the cars themselves. Instead the answer lay in the outlook of those enthusiasts—those "creative eccentrics," as Thompson would have put it—who decided to place an emotional premium on certain cars. I have chosen my words carefully here, because although a monetary premium may well follow in due course, and often does, less tangible forms of value are what really help to turn a neglectable into a collectable. Some Beetle enthusiasts were drawn by the countercultural image of their cars, while others appreciated their easy customizability. First-generation Mustang enthusiasts valued the ease with which their cars could be modified, too, and they also saw something pure in that first pony car that was absent both in its heavier descendants of the early 1970s and in its considerably downsized name-sakes of the mid- to late 1970s. For other groups of gearheads, the operative emotion was nostalgia, either for one's own past—and youth—or for a more distant era. Street rodders looked back wistfully at the California hot-rod scene of the 1930s and 1940s, while street-machine enthusiasts waxed nostalgic for the stylish and powerful cars of the 1950s and 1960s, antique hobbyists for the brass and polished wood of the 1900s and 1910s, and so on.

For others the operative emotion was instead a potent mix of nostalgia and despair that centered on *styling* and *feel*. Consider the following, penned by the automotive journalist Ken Purdy back in 1949:

> Once upon a time the American people thought of the automobile as an instru-ment of sensuous pleasure. Once upon a time, indeed, it was just that, and many a man alive still remembers the day. Aye, many a man crossing a city street that is a turgid river of jelly-bodied clunkers too clumsy to get out of their own way remembers a day when he ripped down a country road in a canary yellow, bucket-seated Mercer, master of all he surveyed, high-riding, able to see where he was going, with a wheel in his hands that really steered the car instead of shyly suggesting that it change direction. . . . It's all past now.

Consequently, he concluded, "We have raised a generation of Americans who have been cheated out of one of life's most important pleasures: the joy of driving a light, fast, safe and supple automobile."[3] Three years later, Eugene Jaderquist concurred in *True's Automotive Yearbook*. The typical par-ticipant in the growing hobby of classic-car restoration, he explained, was "very contemptuous of Detroit stock cars today, a firm subscriber to the theory that 'they don't build them like that anymore.'"[4]

Jaderquist and Purdy were by no means alone. Although their deroga-tory statements regarding early postwar cars surely would have been unwelcome at a special-interest restoration show two decades later, their basic premise—that the cars of yesterday were superior to those of today—lived on. The key, of course, was that with the passage of time, "yesterday" and "today" steadily assumed new meanings, and the reasons justifying a preference for the former gradually shifted. For Jaderquist and Purdy, postwar cars lacked the sensuous curves and graceful poise of the prewar classics. But for a Carter-era street machine enthusiast, the newer vehicles of his time were undesirable because they were deemed to be slower, uglier, difficult to modify legally, and more costly than those of the 1960s. Hence, among other things, the emergent collectability of 1960s Camaros, Chevelles, and other pony and muscle cars in the mid- to late 1970s—and their skyrocketing values by the late 1990s and early 2000s. Hence as well the notion that 1972 somehow marked a watershed: for certain enthusiasts, it did.

On the other hand, some gearheads ultimately came to view those later, Carter-era cars in a more favorable light. Compared with the smaller vehicles of the Reagan, Bush, and Clinton years, with their plastic body panels, complex electronics, and four-cylinder engines, the V-8 cars of the mid- to late 1970s were not only more appealing than contemporary vehicles but also more affordable than the coveted cars of the 1960s. Our turn-of-the-century hot rodder was an excellent case in point: he actually did not share the dominant attitude that he described, for he was among those who were beginning to work on and embrace the V-8 cars built after 1972—and he was unapologetic about it, too. He still favored the old over the new, though what was "old" and what was "new" had clearly shifted.[5]

But a favorable attitude toward that which was old—whatever "old" happened to mean to a given group of gearheads at a given point in time— did not necessarily go hand in hand with a blanket rejection of the new. Hot rodders long embraced any "new and improved" technology that might have given them an advantage on the lakes or the drag strip. Likewise, street rodders baked new technologies into their old cars from the very beginning, rejecting mechanical brakes for hydraulic, manual windows and seats for power, and the blistering heat of an old closed car for the comfort of retrofitted air-conditioning. Many automotive journalists also published open critiques of those who romanticized the automotive past by reminding them that muscle cars were heavy and handled poorly or that most of the cars of the 1950s and 1960s were in fact quantifiably slower than those of the 1970s and 1980s. "Our fantasies about the wonderful cars of yesteryear are just that—bloated dreams," as Car and Driver's Brock Yates put it in 1977. "Today's automobiles are taking a bum rap for being feeble and slow."[6] Yates's colleague Martin Toohey by and large agreed, although he added an important catch. "Compared with today's automotive offerings," he explained, "even the strongest of the classics probably would fail a comparison test's rigid statistical discipline. Quarter-mile times, 0–60 performance and braking distances would, of course, lose to the measuring machines—but alas, there are no love meters to record man's emotional relationship with his car. And therein lies the reason for our regular editorial pageantry on behalf of yesterday's automobiles."[7]

Therein lay as well the inspiration for salvage-yard treasure hunters, barn-find seekers, swap-meet scroungers, and those who cultivated a preference for patina over perfection. Therein also lay what motivated the outspoken opponents of restrictive zoning, ex post facto NIMBYism, and any of a number of old-car scrappage schemes. It was about *feel*, and *love*, and a sense of *emotional connection* with the automotive past—a past that cried out not for sterile preservation at the hands of the few but for genuine enjoyment at the hands of the many.

Notes

SCI	*Sports Cars Illustrated*
SEMA-DC	"EPA: Clunker" group, Specialty Equipment Market Association records, Washington, D.C.
SN	*SEMA News*
SPT	*St. Petersburg Times*
SS	*Street Scene*
TA	*The Automobile*
U-ARB	unarchived ARB records
VWT	*VW Trends*
WP	*Washington Post*
WSJ	*Wall Street Journal*
WT	*Washington Times*
WTG	*Telegram and Gazette* (Worcester, Mass.)

Introduction

1. Allen Sinsheimer, "We Tear 'Em Up and Sell the Pieces," *MA*, 26 April 1917, 5–9 (quote on 5). The used-car trade: Steven M. Gelber, *Horse Trading in the Age of Cars: Men in the Marketplace* (Baltimore, 2008). Automobiles and the scrap-metal trade: Carl Zimring, *Cash for Your Trash: Scrap Recycling in America* (New Brunswick, N.J., 2005); Zimring, "Neon, Junk, and Ruined Landscape: Competing Visions of America's Roadsides and the Highway Beautification Act of 1965," in *The World beyond the Windshield: Roads and Landscapes in the United States and Europe*, ed. Christof Mauch and Thomas Zeller (Athens, Ohio, 2008), 94–107; Zimring, "The Complex Environmental Legacy of the Automobile Shredder," *Technology and Culture* 52 (July 2011): 523–47; Tom McCarthy, *Auto Mania: Cars, Consumers, and the Environment* (New Haven, Conn., 2007); Adam Minter, *Junkyard Planet: Travels in the Billion-Dollar Trash Trade* (New York, 2013). The pillars of automobility: James J. Flink, *America Adopts the Automobile, 1895–1910* (Cambridge, Mass., 1970); Flink, *The Automobile Age* (Cambridge, Mass., 1988); John B. Rae, *The American Automobile: A Brief History* (Chicago, 1965); Rae, *The Road and the Car in American Life* (Cambridge, Mass., 1971); John A. Jakle and Keith A. Sculle, *The Gas Station in America* (Baltimore, 1994); Jakle and Sculle, *Motoring: The Highway Experience in America* (Athens, Ga., 2008); Clay McShane, *Down the Asphalt Path: The Automobile and the American City* (New York, 1994); Rudi Volti, *Cars and Culture: The Life Story of a Technology* (Westport, Conn., 2004); Kevin L. Borg, *Auto Mechanics: Technology and Expertise in Twentieth-Century America* (Baltimore, 2007); Cotten Seiler, *Republic of Drivers: A Cultural History of Automobility in America* (Chicago, 2008); John Heitmann, *The Automobile and American Life* (Jefferson, N.C., 2009); Joseph J. Corn, *User Unfriendly: Consumer Struggles with Personal Technologies, from Clocks and Sewing Machines to Cars and Computers* (Baltimore, 2011).

2. "Estimated Number of U.S. Passenger Cars Scrapped," *AI*, 27 February 1932, 287.

3. Sewage systems, garbage transfer stations, landfills, incinerators, recycling centers, and other elements of the waste-management infrastructure: Kevin Lynch with Michael Southworth, *Wasting Away: An Exploration of Waste* (San Francisco, 1990); Joel A. Tarr, *The Search for the Ultimate Sink: Urban Pollution in Historical Perspective* (Akron, Ohio, 1996); Susan Strasser, *Waste and Want: A Social History of Trash* (New York, 1999); Martin V. Melosi, *The Sanitary City: Environmental Services in Urban Amer-*

ica from Colonial Times to the Present (Baltimore, 2000); Melosi, Garbage in the Cities: Refuse, Reform, and the Environment, rev. ed. (Pittsburgh, 2005); William Rathje and Cullen Murphy, Rubbish! The Archaeology of Garbage, rev. ed. (Tucson, 2001); Elizabeth Royte, Garbage Land: On the Secret Trail of Trash (New York, 2005); Giles Slade, Made to Break: Technology and Obsolescence in America (Cambridge, Mass., 2006); McCarthy, Auto Mania, esp. 84–85, which compares American perspectives on the disposal of automotive waste and human excrement; Zimring, Cash for Your Trash; Edward Humes, Garbology: Our Dirty Love Affair with Trash (New York, 2012). International perspectives: Donald Reid, Paris Sewers and Sewermen: Realities and Representations (Cambridge, Mass., 1991); Stephen Halliday, The Great Stink of London: Sir Joseph Bazalgette and the Cleansing of the Victorian Metropolis (Stroud, Gloucestershire, U.K., 1999); Finn Arne Jørgensen, Making a Green Machine: The Infrastructure of Beverage Container Recycling (New Brunswick, N.J., 2011); Raymond G. Stokes, Roman Köster, and Stephen C. Sambrook, The Business of Waste: Great Britain and Germany, 1945 to the Present (New York, 2013). LULUs and NIMBYism: Robert H. Nelson, Zoning and Property Rights: An Analysis of the American System of Land-Use Regulation (Cambridge, Mass., 1977); William A. Fischel, The Economics of Zoning Laws: A Property Rights Approach to American Land Use Control (Baltimore, 1985); John O'Looney, Economic Development and Environmental Control: Balancing Business and Community in an Age of NIMBYs and LULUs (Westport, Conn., 1995); Herbert Inhaber, Slaying the N.I.M.B.Y. Dragon (New Brunswick, N.J., 1998); Gregory McAvoy, Controlling Technocracy: Citizen Rationality and the NIMBY Syndrome (Washington, D.C., 1999); Daniel W. Bromley, "Property Rights and Land Use Conflicts: Reconciling Myth and Reality," in Economics and Contemporary Land Use Policy: Development and Conservation at the Rural-Urban Fringe, ed. Robert J. Johnston and Stephen K. Swallow (Washington, D.C., 2006), 38–51.

4. Ancient, early modern, and more recent habits and methods of disposing of domestic, industrial, and hazardous wastes: D. J. Keene, "Rubbish in Medieval Towns," in Environmental Archeology in the Urban Context, ed. A. R. Hall and H. K. Kenward (York, U.K., 1982), 26–30; Dolly Jørgensen, "Cooperative Sanitation: Managing Streets and Gutters in Late Medieval England and Scandinavia," Technology and Culture 49 (July 2008): 547–67; Ronald E. Zupko and Robert A. Laures, Straws in the Wind: Medieval Urban Environmental Law—The Case of Northern Italy (Boulder, Colo., 1996), esp. chaps. 3–5; Susan J. Kleinberg, "Technology and Women's Work: The Lives of Working Class Women in Pittsburgh, 1870–1900," in Dynamos and Virgins Revisited: Women and Technological Change in History, ed. M. M. Trescott (Lanham, Md., 1979), 185–204; William Cronon, Nature's Metropolis: Chicago and the Great West (New York, 1991), esp. chap. 5; Tarr, The Search for the Ultimate Sink; Harold L. Platt, Shock Cities: The Environmental Transformation and Reform of Manchester and Chicago (Chicago, 2005); Jacob Darwin Hamblin, Poison in the Well: Radioactive Waste in the Oceans at the Dawn of the Nuclear Age (New Brunswick, N.J., 2009).

5. Flink, America Adopts the Automobile, 55–62; Rae, American Automobile, chap. 3; Heitmann, Automobile and American Life, chap. 1; Volti, Cars and Culture, chap. 2.

6. McCarthy, Auto Mania, chap. 8; Zimring, "Neon, Junk, and Ruined Landscape"; Zimring, "Complex Environmental Legacy"; Zimring, Cash for Your Trash, chap. 5. See also below, chaps. 1 and 5.

7. In popular music, witness Jim Croce's "Bad, Bad Leroy Brown" (1973), the Zac Brown Band's "Junkyard" (2005), and Left Lane Cruiser's album *Junkyard Speed Ball* (2011). In television, consider *Knight Rider*'s season 2 "Blind Spot" (1983) and season 3 "Junk Yard Dog" (1985), *Law and Order: Special Victims Unit*'s season 6 "Parts" (2005), *Dexter*'s season 1 "Return to Sender" (2006), and the search for a junkyard dog (and discovery of a junkyard cat) in *It's Always Sunny in Philadelphia*'s season 3 "Bums" (2007). In literature, see Russell Hoban, *The Mouse and His Child* (New York, 1967); Janisse Ray, *Ecology of a Cracker Childhood* (Minneapolis, 1999); Michael Coles, "Smash Palace— Model Four Hundred," *Fourth Genre* 8 (2006): 139–41; and Steven Church, "Danger Boys," *River Teeth* 10 (Fall 2008–Spring 2009): 269–78. On the debate within the business over "junkyard," "salvage yard," and other alternatives, see Zimring, "Neon, Junk, and Ruined Landscape," 102–3; and Zimring, *Cash for Your Trash*, 21; see also below, chap. 4.

8. The history of local ordinances over the last fifty years bears this out. See below, chap. 5; Carol A. Marbin, "Neighbors Upset Over Junkyard," *SPT*, 29 June 1990; Julie E. Gerrish, "Student Goes Head-On After Junk Cars," *WTG*, 16 May 1995; Amy Jeter, "Darby Twp. Tires of 'Junkyard' on Hook Road," *PI*, 25 February 2000; Phil Trexler, "City Clears 'Junkyard' Eyesore," *Akron Beacon Journal*, 17 October 2007; and Jonah Owen Lamb, "Block Resembles Junkyard," *Merced Sun-Star* (California), 6 August 2009.

9. Some examples: Dan Fisher, "There May Be a Lot of Recycled Detroit in That Japanese Import," *LAT*, 23 April 1972; Suzanne DeChillo, "Making Much Do from 'Nothing,'" *NYT*, 18 July 1982; Paul Hemp, "Junkyards Polish Their Rusty Image," *NYT*, 30 August 1983; James R. Chiles, "The Great American Junkyard," *Smithsonian*, March 1985, 52–63; Fred McMorrow, "About Long Island," *NYT*, 8 February 1987; George P. Blumberg, "Junkyards Discard an Image, and the Scary Dogs, Too," *NYT*, 23 October 2002; Marty Clear, "New Breed of Used-Parts Yards," *SPT*, 24 December 2004.

10. "Gearhead" is a common synonym for "automotive enthusiast." For a discussion of automotive enthusiasm in its many forms, see below, chap. 2; see also Dale Dannefer, "Neither Socialization nor Recruitment: The Avocational Careers of Old-Car Enthusiasts," *Social Forces* 60 (December 1981): 395–413; H. F. Moorhouse, *Driving Ambitions: A Social Analysis of the American Hot Rod Enthusiasm* (New York, 1991); Robert C. Post, *High Performance: The Culture and Technology of Drag Racing, 1950–1990* (Baltimore, 1994); David N. Lucsko, *The Business of Speed: The Hot Rod Industry in America, 1915–1990* (Baltimore, 2008); and Jakle and Sculle, *Motoring*, 4, 225–26. Groups similar in everything but the object(s) of their enthusiasm are discussed—as "enthusiasts," "hobbyists," or "hedonists"—in Christina Lindsay, "From the Shadows: Users as Designers, Producers, Marketers, Distributors, and Technical Support," in *How Users Matter: The Co-Construction of Users and Technologies*, ed. Nelly Oudshoorn and Trevor Pinch (Cambridge, Mass., 2003), 29–50; John Davis, "Going Analog: Vinylphiles and the Consumption of the 'Obsolete' Vinyl Record," in *Residual Media*, ed. Charles R. Acland (Minneapolis, 2007), 222–36; Rachel Maines, *Hedonizing Technologies: Paths to Pleasure in Hobbies and Leisure* (Baltimore, 2009); and Kieran Downes, "'Perfect Sound

Forever': Innovation, Aesthetics, and the Re-making of Compact Disc Playback," *Technology and Culture* 51 (2010): 305–31.

11. McCarthy, *Auto Mania*, chap. 5.

12. Dean Shipley, "Schulte's Auto Wrecking," *C&P*, March 1996, 20; see also below, chap. 3.

13. John Robertson, "Rony's Auto Sales," *C&P*, July 1993, 36. Tony's father, who launched the family's salvage business in 1959, was named Rony.

14. Barry Penfound, "Bob's Auto Wrecking and Recovery, Inc., Milan, Ohio," *C&P*, July 2005, 51–52.

15. See below, chap. 4. The specific examples cited here include "Junkyard Jamboree," *HRM*, February 1977, 70–74, 76–83; Willard C. Poole, "Old Car Treasure Hunt Is On!" *Cars*, May 1953, 36–37, 79; Jim Losee, "Low Dollar Suspension Parts Guide," *PHR*, March 1984, 46–48, Losee, "Small-Block Chevy Engine Spotter's Guide," *PHR*, May 1984, 36–39, Losee, "Budget GM Engine Parts Shopping," *PHR*, June 1984, 62–65, all part of *PHR*'s "Salvage Yard Treasure Hunt" series; John Thawley, "Wreck-Connaissance," *R&C*, May 1974, 24–26, 37–39; Cam Benty, "The Salvage Yard Maze," *PHR*, August 1982, 76–79; John Lee, "Sullivan Salvage," *C&P*, December 1991, 36–38; and the long-running *HRM* series "Junkyard Jewel," including Steve Dulcich, "400 HP 318," April 2003, 76–82; Marlan Davis, "Pontiac 400," October 2006, 110–14, 116; and Christopher Campbell, "Ford 4.6L 2V," April 2008, 104–8.

16. On swap meets, see below, chap. 4. On Hershey, see Menno Duerksen, "We've Been to Hershey '70," *C&P*, November 1970, 58–67. On creative repurposing, see below, chap. 2. On junkyard calendars, posters, and videos, as well as barn finds and the sensory experience of wandering through a salvage yard, see below, chap. 4.

17. See below, chaps. 5–6.

18. Steve Magnante, "Doin' the Junkyard Crawl," *HRM*, August 1999, 44 (quote).

19. Although it focuses primarily on the United States, much of what *Junkyards, Gearheads, and Rust* has to say about these phenomena applies as well to other parts of the world—especially Canada but also Western Europe, Australasia, parts of Latin America, and pockets elsewhere.

20. McCarthy, *Auto Mania*, esp. chap. 5; McCarthy, "Henry Ford, Industrial Ecologist or Industrial Conservationist? Waste Reduction and Recycling at the Rouge," *Michigan Historical Review* 27 (Fall 2001): 53–90; Zimring, "Neon, Junk, and Ruined Landscape"; Zimring, "Complex Environmental Legacy"; Zimring, *Cash for Your Trash*.

21. On the used-parts business, see below, chaps. 1 and 3.

22. On hot rodders and street rodders, see Moorhouse, *Driving Ambitions*; Post, *High Performance*; and Lucsko, *Business of Speed*. On customs, see John DeWitt, *Cool Cars, High Art: The Rise of Kustom Kulture* (Jackson, Miss., 2002); Brenda Jo Bright, "Mexican American Low Riders: An Anthropological Approach to Popular Culture" (Ph.D. diss., Rice University, 1994); and Bright, *Customized: Art Inspired by Hot Rods, Low Riders, and American Car Culture* (New York, 2000). On sports cars and imports, see Jeremy Kinney, "Racing on Runways: The Strategic Air Command and Sports Car Racing in the 1950s," *ICON* 19 (2013): 193–215. Restoration hobbyists have yet to attract much scholarly notice; for an introduction, see below, chap. 2.

23. James H. Moloney, *Encyclopedia of American Cars, 1930 to 1942* (Sarasota, Fla., 1977), 226; Terry Shuler with Griffith Borgeson and Jerry Sloniger, *The Origin and Evolution of the Volkswagen Beetle* (Princeton, N.J., 1985), 84–85; "1970 Challenger Big Block Production," table, www.challengerspecs.com/1970challenger_productionnumbers.htm (accessed 21 October 2014).

24. Douglas Harper, *Working Knowledge: Skill and Community in a Small Shop* (Berkeley, Calif., 1987); Strasser, *Waste and Want*, esp. 10–12 and chap. 1. See also Miles Orvell, *The Real Thing: Imitation and Authenticity in American Culture, 1880–1940* (Chapel Hill, N.C., 1989), esp. 287–99.

25. David Edgerton, *The Shock of the Old: Technology and Global History since 1900* (New York, 2007).

26. Borg, *Auto Mechanics*; Gelber, *Horse Trading*.

27. Vance Packard, *The Hidden Persuaders* (New York, 1957); Packard, *The Status Seekers* (New York, 1959); Packard, *The Waste Makers* (New York, 1960); John Keats, *The Insolent Chariots* (Philadelphia, 1958); John A. Kouwenhoven, *The Beer Can by the Highway: Essays on What's "American" about America* (Garden City, N.Y., 1961); Peter Blake, *God's Own Junkyard: The Planned Deterioration of America's Landscape* (New York, 1964); Lynch, *Wasting Away*.

28. William McDonough and Michael Braungart, *Cradle to Cradle: Remaking the Way We Make Things* (New York, 2002); Slade, *Made to Break*; Royte, *Garbage Land*; Humes, *Garbology*; Minter, *Junkyard Planet*; *Addicted to Plastic: The Rise and Demise of a Modern Miracle*, DVD directed by Ian Connacher, Bullfrog Films, 2007.

29. Rathje and Murphy, *Rubbish*; Michael Thompson, *Rubbish Theory: The Creation and Destruction of Value* (New York, 1979); Charles R. Acland, "Introduction: Residual Media," xiii–xxvii, Davis, "Going Analog," 222–36, Michelle Henning, "New Lamps for Old: Photography, Obsolescence, and Social Change," 48–65, Lisa Parks, "Falling Apart: Electronics Salvaging and the Global Media Economy," 32–47, Hillegonda C. Rietveld, "The Residual Soul Sonic Force of the 12-Inch Dance Single," 97–114, and Jonathan Sterne, "Out with the Trash: On the Future of New Media," 16–31, all in Acland, *Residual Media*; Melosi, *The Sanitary City*; Melosi, *Garbage in the Cities*; Strasser, *Waste and Want*; McCarthy, *Auto Mania*, chap. 5; McCarthy, "Henry Ford, Industrial Ecologist or Industrial Conservationist?"; Zimring, "Neon, Junk, and Ruined Landscape"; Zimring, "Complex Environmental Legacy"; Zimring, *Cash for Your Trash*; Stokes, Köster, and Sambrook, *The Business of Waste*; Jørgensen, *Making a Green Machine*.

30. Downes, "Perfect Sound Forever"; Davis, "Going Analog"; Rietveld, "Residual Soul Sonic Force." See also Leo Marx, "Does Improved Technology Mean Progress? Understanding the Historical Distinction between Two Contradictory Concepts of Progress Helps Explain the Current Disenchantment with Technology," *Technology Review*, January 1987, 32–41, 71.

31. One need look no further than the veritable marketplace stampede toward the qualitatively inferior mp3 format over the last fifteen years—or the equally remarkable phenomenon of the annual queue at the Apple Store—to realize just how exceptional these old-media enthusiasts are.

32. Centered as it often is on moments of novelty, much of the survey literature on the history of the automobile suggests the same conclusion. See Rae, *American Automo-*

bile; Frank Donovan, *Wheels for a Nation: How America Fell in Love with the Automobile and Lived Happily Ever After . . . Well, Almost* (New York, 1965); Richard Crabb, *Birth of a Giant: The Men and Machines That Gave America the Motorcar* (Philadelphia, 1969); Jean-Pierre Bardou, Jean-Jacques Chanaron, Patrick Fridenson, and James M. Laux, *The Automobile Revolution: The Impact of an Industry*, ed. and trans. James M. Laux (Chapel Hill, [1977] 1982); and Flink, *Automobile Age*.

33. Thompson, *Rubbish Theory*, 7–17; see also below, chaps. 5–6.

34. For the former, see Ken W. Purdy, *The Kings of the Road* (Boston, 1949); Warren Weith, "Cars," *C&D*, July 1964, 25, 27; Weith, "Life, Lessons, and Old Cars," *C&D*, January 1980, 22; and Brock Yates, *The Decline and Fall of the American Automobile Industry* (New York, 1983). For the latter, see Brock Yates, untitled editorial, *C&D*, March 1974, 21; Yates, "A Closer Look at Today versus the Good Old Days," *C&D*, December 1977, 22; and Warren Weith, "Fordor Then and Now," *C&D*, November 1980, 28.

1. The Automotive Salvage Business in America

1. Michelle Krebs, "New Ford Division to Focus on Recycling of Auto Parts," *NYT*, 27 April 1999; Krebs, "Ford's Recycling Buy May Shorten Repair Waiting Time," *WT*, 28 May 1999; "Nasser's 'Better Idea,'" *AN*, 10 May 1999, 12 (quote); Brian Akre, "Ford Revving Up to Be an Auto Recycling Giant," *Toronto Star*, 15 May 1999; "Ford Is Set to Unveil 'Internet Junkyard' for Selling Used Parts," *WSJ*, 26 April 1999; Bryce Hoffman, "Ford Finding Treasure in Trash," *AN*, 3 January 2000, 16; Joseph B. White, "Ford 'Greenleaf' Program Is Investigated in Florida over Disclosure Allegations," *WSJ*, 9 May 2001; Robert Sherefkin, "Lawsuit May Tangle Ford's Greenleaf Sale," *AN*, 22 July 2002, 8; "Ford Sells GreenLeaf Recycling Chain," *AN*, 14 July 2003, 26. On Nasser's departure, see Joseph B. White, "Ford's Jacques Nasser Is Ousted as CEO," *WSJ*, 30 October 2001. On Ford's 1930s–1940s program, see Tom McCarthy, "Henry Ford, Industrial Ecologist or Industrial Conservationist? Waste Reduction and Recycling at the Rouge," *Michigan Historical Review* 27 (Fall 2001): 53–90; McCarthy, "Henry Ford, Industrial Conservationist? Take-Back, Waste Reduction and Recycling at the Rouge," *Progress in Industrial Archaeology* 3, no. 4 (2006): 302–28; Gerald Perschbacher, "Scrap Program Helped Stop Bleeding," *OCW*, 26 March 2009, 22; and Charles E. Sorensen, *My Forty Years with Ford* (Detroit, [1956] 2006), 174–75.

2. Kevin Leininger, "Pull-A-Part Is Not Junkyard," *News-Sentinel* (Fort Wayne, Ind.), 3 November 2008; Brent Hunsberger, "Profits in Parts," *Oregonian* (Portland), 6 February 2005; Robert D. Hof, "Turning Rust into Gold," *BW*, 10 June 2002, 102.

3. Peter B. Nugent, "Auto Recycler Defies Stereotype of Messy Junkyard," *WTG*, 16 December 1992; Jeff Haynes, "Yesterday's Junkyards Are Today's High-Tech Recyclers," *Jacksonville Business Journal*, 4 September 1998; Rachel Melcer, "Need a Used Manifold? Car-Part.com Lets You Scrap Trip to Junkyard," *Business Courier* (Cincinnati), 24 November 2000; Frank Hotchkiss, "Buy Auto Parts, New, Used over Internet," *WT*, 12 November 1999; "Planet Salvage," *CHP*, December 2000, 16.

4. John Langley, "Automated Car Recycling Takes Place at Unique Plant," *WT*, 12 July 1991; Maryann Keller, "Europe Recycles," *AI*, November 1991, 9; Jean V. Owen, "Environmentally Conscious Manufacturing," *Manufacturing Engineering*, October 1993, http://proquest.umi.com/pqdweb?did=1209197&sid=1&Fmt=3&clientId=1997&RQT

=309&VName=PQD (accessed 21 April 2010); Bill Tuckey, "Car Makers Plan for Life after Death," *The Age* (Melbourne, Australia), 25 March 1994; Dennis Simanaitis, "Tech Tidbits," *R&T*, June 1999, 181–83; "Ford's Tired Recycling," *AI*, May 1994, 17; "Junkyards Join Recycling Study," *AI*, 21 March 1994, 2.

5. Philip Marcelo, "Auto Salvage Business Changes Routine to Keep Up with Times," *Providence Journal*, 11 January 2007.

6. John L. Moore, "Used Auto Parts Dealer Driven by Latest Trends to Grow," *Northeast Pennsylvania Business Journal*, 1 October 1998, 42; Phil Skinner, "Renegade Wrecker, Law Abiding Citizen," *C&P*, December 1999, 56–57, 59.

7. Carl Zimring, *Cash for Your Trash: Scrap Recycling in America* (New Brunswick, N.J., 2005), chap. 2.

8. See Angelia Riveira, "A Tale of Two Markets: The People and Culture of American Flea Markets" (M.A. thesis, Auburn University, 2013), chap. 1; Things the Motorist Wants to Know, *Motor*, June 1908, 60; "Motor Junk Is Sold at Fair," *MA*, 3 June 1915, 28; and below, chaps. 2 and 4.

9. "One Use for Abandoned Cars," *TA*, 28 May 1908, 758; "Junk Man Rides in Hash-Mobile," *MA*, 11 May 1916, 13. See also Kevin L. Borg, *Auto Mechanics: Technology and Expertise in Twentieth-Century America* (Baltimore, 2007), chap. 2; and below, chap. 2.

10. "What Becomes of Out of Date Cars?" *HA*, 31 May 1905, 600; M. C. Krarup, "Bargains in Second-Hand Motor Cars," *Motor*, June 1905, 34, 78, 80, 82, 84, 85, 88; "One Use for Abandoned Cars"; Things the Motorist Wants to Know; "Where the Old Cars Go," *TA*, 20 May 1909, 841. See also below, chap. 4.

11. Cugnot: Rudolph E. Anderson, *The Story of the American Automobile: Highlights and Sidelights* (Washington, D.C., 1950), 8–9; Kit Foster, "1771 Cugnot Fardier a Vapeur," *Autoweek*, 1 April 2001, http://autoweek.com/article/car-news/1771-cugnot-fardier-vapeur-mother-all-motorcars (accessed 5 January 2015). Dudgeon: John Bentley, *Oldtime Steam Cars* (New York, 1953), 20–21.

12. Clay McShane, *Down the Asphalt Path: The Automobile and the American City* (New York, 1994), esp. chaps. 3, 5, and 9; Peter Norton, "Street Rivals: Jaywalking and the Invention of the Motor Age Street," *Technology and Culture* 48 (April 2007): 331–59; Norton, *Fighting Traffic: The Dawn of the Motor Age in the American City* (Cambridge, Mass., 2008).

13. Advocates for Highway Safety, "Motor Vehicle Traffic Fatalities and Fatality Rate: 1899–2003 (Based on Historical NHTSA and FHWA Data)," table, www.saferoads.org/federal/2004/TrafficFatalities1899-2003.pdf (accessed 5 January 2015); Federal Highway Administration, U.S. Department of Transportation, "State Motor Vehicle Registrations, by Years, 1900–1995," table, www.fhwa.dot.gov/ohim/summary95/section2.html (accessed 5 January 2015).

14. Prior to the 1970s, only on occasion did "totaled" appear in the automotive press without quotation marks, suggesting that the term was not yet widely used even though the practice itself was as old as the automobile. By the 1990s, it normally appeared without quotation marks, indicating broader use. See *Oxford English Dictionary*, www.oed.com, s.v. "total" (accessed 5 January 2015); see also George Finneran, "Major Conservatives," *Restyle Your Car*, 1952 annual, 74 ("total" in quotes); Eric Rickman,

"Build a Windsor-Merc," *HRM*, December 1956, 22 ("total" in quotes); Tex Smith, "A Guide to Best Buys in . . . USED PARTS," *PHR*, August 1967, 74 ("totaled out" in quotes); Jay Jones, "2002 Transformed," *EC*, July–August 1991, 76 ("totaled" without quotes); Robert K. Smith, "Sixty-Six Streeter," *HVW*, February 1998, 49 ("total" without quotes). On the early car-insurance business, see Darwin S. Hatch, "The Plain English of Motor Car Insurance," *MA*, 15 April 1915, 5–10.

15. Robin Damon, "Too Rapid Depreciation," *HA*, 18 February 1903, 272–73; "What Becomes of Out of Date Cars?"; "Second Hand Cars," *HA*, 9 May 1906, 658; Herbert L. Towle, "When Is a Car Worn Out? What Percentage of First Cost Should Be Written Off Annually for Depreciation? The Noise Produced As an Indication of the End of a Car's Usefulness," *Motor*, April 1910, 72 (quotes); James J. Flink, *America Adopts the Automobile, 1895–1910* (Cambridge, Mass., 1970), chap. 1; Borg, *Auto Mechanics*, chaps. 1–2.

16. "Old Autocar Recalls Ancient History," *TA*, 16 April 1908, 552; Towle, "When Is a Car Worn Out?"; The Reader's Clearing House, *MA*, 11 September 1913, 36.

17. "What Becomes of Out of Date Cars"; Krarup, "Bargains in Second-Hand Motor Cars"; "What Becomes of the Old Automobiles?" *HA*, 27 March 1907, 438; "One Use for Abandoned Cars" (quote); "Where the Old Cars Go."

18. Allen Sinsheimer, "We Tear 'Em Up and Sell the Pieces," *MA*, 26 April 1917, 5–9.

19. Ibid., 7–8 (on the Auto Salvage and Parts House, quote on 7); G. N. Georgano, ed., *The Complete Encyclopedia of Motorcars, 1885 to the Present*, 2nd ed. (New York, 1973), 558 (on the Pope-Toledo).

20. Sinsheimer, "We Tear 'Em Up," 7–8.

21. Ibid., 8.

22. Ibid., 6 (quote); Auto Salvage Company advertisements, *MA* and *TA*, 1915–1920.

23. Sinsheimer, "We Tear 'Em Up," 7 (quote). The Auto Salvage and Wrecking Company of Oklahoma City used a slogan similar to that of the Auto Wrecking Company: "Always Tearing 'Em Up and Selling the Pieces" (advertisements, *MA*, 1919–1921).

24. Sinsheimer, "We Tear 'Em Up," 6–7. This card system, designed to mitigate shop-floor chaos, is reminiscent of the list system developed at Baldwin Locomotive in the nineteenth century. See John K. Brown, *The Baldwin Locomotive Works, 1831–1915: A Study in American Industrial Practice* (Baltimore, 1995), chap. 4.

25. Prompt shipping: Spardacine Auto Wrecking Company advertisement, *TA*, 28 December 1916, 93; Sinsheimer, "We Tear 'Em Up," 7; Bloomington Auto Salvage Company advertisement, *MA*, 28 June 1917, 148. Range of stock: Wichita Auto Wrecking Company advertisements, *MA*, 1918–1924 (advertisements for this firm only began to appear in 1918, though its operations date back well before World War I); Sinsheimer, "We Tear 'Em Up." Money-back guarantees: Bloomington Auto Salvage Company advertisement; Auto Salvage Company advertisement, *MA*, 29 June 1916, 113; Sinsheimer, "We Tear 'Em Up." Overall savings: Auto Wrecking Company advertisements, *TA*, *MA*, and *AI*, 1915–1923; Bloomington Auto Salvage Company advertisement; Spardacine Auto Wrecking Company advertisement; Auto Salvage Company advertisements, *MA* and *TA*, 1915–1920. "Salvage" versus "junk": Auto Salvage Company advertisement, *MA*, 29 June 1916, 113 (embracing "salvage"); Sinsheimer, "We Tear 'Em Up," 8 (embracing

"junk"). See also below, chap. 4; Carl Zimring, "Neon, Junk, and Ruined Landscape: Competing Visions of America's Roadsides and the Highway Beautification Act of 1965," in *The World beyond the Windshield: Roads and Landscapes in the United States and Europe*, ed. Christof Mauch and Thomas Zeller (Athens, Ohio, 2008), 102–3; and Zimring, *Cash for Your Trash*, 21.

26. W. F. Bradley, "Saving Trucks at the Front in France," *TA*, 14 January 1915, 51–55, 59 (quote on 53).

27. W. F. Bradley, "Rejuvenating Wrecked Cars," *TA*, 29 June 1916, 1149–53 (quote on 1150); Bradley, "French War Repair Shops Models of Systematism," *MA*, 29 June 1916, 20–21, 36.

28. German depot: "Photographs Reaching This Country on the Deutschland, German's Super-Submersible, Show Kaiser's Motor Vehicle Toll to Be Excessive," *MA*, 20 July 1916, 19; "Extra," *Commercial Vehicle*, 1 August 1916, 25 (both of these sources, likely from the same author, report that the depot was located at "Gempelhof" in Berlin, surely a misspelling of "Tempelhof" in the newswire). American depot: W. F. Bradley, "M. T. C. Salvage Park in France, Part I," *AI–TA*, 17 April 1919, 860–63; Bradley, "M. T. C. Salvage Park in France, Part II," *AI–TA*, 24 April 1919, 902–5.

29. See, for example, Newton Auto Salvage Company advertisement, *MA*, 26 June 1919, 151; Illinois Auto Parts Company advertisement, *MA*, 30 June 1921, 129; Indiana Auto Parts and Tire Company advertisement, *MA*, 29 June 1922, 126; United Auto Wreckers advertisements, *MA*, 1926–1927; and Douglas Auto Parts Company advertisements, *MA*, 1924–1925; Sam Coraz Auto Parts and Tire Company advertisement, *MA*, 25 December 1919; Warshawsky and Company advertisement, *MA*, 29 June 1922, 125.

30. Daniel Strohl, "Ernest Holmes, a Cadillac, and the Invention of the Tow Truck," *Hemmings Daily*, 5 June 2014, http://blog.hemmings.com/index.php/2014/06/05/ernest-holmes-a-cadillac-and-the-invention-of-the-tow-truck/ (accessed 7 January 2015); Holmes Auto Wreckers advertisement, *MA*, 19 July 1923, 59; "A Home-Made Wrecking Truck That Does the Job," *MA*, 23 September 1920, 26; "Get Extra Business with a Service Truck," *MA*, 14 October 1920, 11; "Scrapping the Ancient Auto," *Literary Digest*, 17 April 1926, 21–22.

31. Motor Salvage Company advertisement, *MA*, 26 December 1918, 103; Fort Dodge Auto Wrecking Company advertisement, *MA*, 26 June 1919, 147; Memphis Auto Parts Company advertisement, *MA*, 26 June 1919, 150; Universal Auto Parts House advertisement, *MA*, 24 June 1920, 112.

32. Precise numbers vary by source (and what each counted as "scrapped"). In 1925, *Motor Age* put the total in 1920 at 273,000, 1923 at 1,010,000, and 1924 at 1,450,000. Seven years later, *Automotive Industries* published data for 1923 through 1931, placing the number scrapped in 1923 at 900,000 and in 1924 at 1,177,300. Within two years, however, *Automotive Industries* had revised its 1924 figure to 1,102,000. Consequently, although it may have happened in 1923, we can only safely say that the nation passed the million-per-year mark by 1924. See Sam Shelton, "Optimism Is Popular Note at New York Show," *MA*, 15 January 1925, 11; "Estimated Number of U.S. Passenger Cars Scrapped," *AI*, 27 February 1932, 287; and "2,350,000 Motor Vehicles Scrapped in 1933," *AI*, 24 February

1934, 217. On the abandoned-car problem of the 1920s, see Tom McCarthy, *Auto Mania: Cars, Consumers, and the Environment* (New Haven, Conn., 2007), 86.

33. Zimring, *Cash for Your Trash*, 116.

34. "'Rosey,' the Junkman, Has Made a Fortune Out of Old Motor Cars," *American Magazine*, September 1924, 69–70 (accessed courtesy Auburn University Libraries Special Collections); "They Go to the Boneyard," *MA*, 20 May 1926, 21; "The Wanderings and Latter Ends of Some Old Cars," *Literary Digest*, 15 October 1921, 52 (quote); The Bone Yard/The Boneyard/Auto Boneyard advertisements, *AI–TA* and *MA*, 1917–1923. See also McCarthy, *Auto Mania*, 84–86.

35. "Map Drive to Save Westchester Beauty," *NYT*, 27 February 1929, 44; "Ways of Controlling Automobile 'Graveyards,'" *American City*, October 1929, 171; McCarthy, *Auto Mania*, 86. Land-use conflicts between open yards and their neighbors have persisted to the present. See below, chap. 5; Zimring, "Neon, Junk, and Ruined Landscape"; and McCarthy, *Auto Mania*, esp. 155.

36. Precise numbers vary. See "Estimated Number of U.S. Passenger Cars Scrapped" and "2,350,000 Motor Vehicles Scrapped in 1933."

37. "The Second Hand Car Problem," *HA*, 5 April 1911, 593; see also Steven M. Gelber, *Horse Trading in the Age of Cars: Men in the Marketplace* (Baltimore, 2008), chap. 4.

38. Speedsters: "Transforming Used Cars into Speedsters Is Profitable," *MA*, 25 June 1914, 36; "He Fits His Cars with Speedster Bodies," *MA*, 24 March 1921, 33 (quote); David N. Lucsko, *The Business of Speed: The Hot Rod Industry in America, 1915–1990* (Baltimore, 2008), chap. 1. Truck conversions: "Putting the Used Car among Dealer's Assets," *HA*, 1 November 1917, 18–19, 88; "Transforming Used Passenger Cars into Trucks," *HA*, 1 March 1918, 29–31. Consignments: "Proposed Solution of the Used Car Problem," *HA*, 4 December 1912, 852. Used-car shows: "Used Car Problems," *MA*, 10 May 1917, 12; "Seattle Dealers' Plan Again Cleans Up Used Car Stocks," *MA*, 17 March 1921, 15.

39. "Used Cars Pile Up as New Sales Grow," *AI–TA*, 26 May 1921, 1128; "Used Cars Still Problem," *AI–TA*, 3 August 1922, 235.

40. "Illinois Dealers Try to Solve Used Car Problems by Organizing," *MA*, 17 June 1920, 27; "Ford Dealers Combine for Used Car Disposal," *AI–TA*, 19 May 1921, 1089; "Inaugurate Raffle Sales of Used Cars in Alabama," *MA*, 14 July 1921, 31; "Rebuilding the Used Car," *MA*, 3 November 1921, 56–57.

41. Fred M. Loomis, "Breaking Even on Used Cars," *MA*, 17 April 1919, 20–21.

42. "Owners Say, 'Scrap Used Trucks,'" *AI*, 5 July 1923, 35; "Used Trucks Yield Profits from Parts," *AI*, 30 August 1923, 449 (also published as "Some Dealers Find Profit in Dismantling Used Trucks," *MA*, 6 September 1923, 40); "Breaks 'Em Up and Sells the Pieces," *MA*, 10 January 1924, 27; "Cars That Should Be Junked," *MA*, 6 November 1924, 56 (quote). Sixty-five years later, similar arguments were used, sans the overt moral disapprobation and naked declarations of self-interest, to sell American consumers on corporate-sponsored old-car scrappage schemes. See below, chap. 6.

43. A. B. Waugh, "Junks Never Leave Here on Four Wheels," *MA*, 12 May 1927, 12–13; Lewis C. Dibble, "Cooperative Salvaging," *MA*, 1 December 1927, 38–40; "Omaha Junking Plan Proves Profitable," *AI*, 23 February 1929, 358. *Motor Age* also published forum discussions on dealer-operated yards in early 1927. See Sam Shelton, "Eliminate

Unmerchantable Car from the Trade," *MA*, 7 April 1927, 12–13; "What Dealers Say about Scrapping Old Cars," *MA*, 14 April 1927, 12–13, 18; and "Many Dealers Favor Scrapping Old Cars," *MA*, 21 April 1927, 12–13, 20.

44. "Dealers Unite in Junking Used Cars," *MA*, 9 June 1927, 13 (quote); H. H. James, "Junking the Old Bones," *MA*, 15 September 1927, 33, 38; "Kansas City Scraps 300 Cars in April," *AI*, 12 May 1928, 753.

45. "Milwaukee Group to Try Wrecking," *MA*, 9 June 1927, 12; L. E. Meyer, "$15,000 to Start a Salvage Yard," *MA*, 19 April 1928, 34–35, 42; "St. Louis Organizes Car Salvage Plant," *AI*, 4 August 1928, 173.

46. "Detroit Dealers to Form Salvage Co.," *MA*, 13 October 1927, 9; Dibble, "Cooperative Salvaging"; "Sioux City Plans Salvage Yard," *AI*, 28 December 1928, 852; "California Dealers Junk 1000 Cars in Eight Weeks," *AI*, 18 August 1928, 245; Leslie Peat, "Dealers Thwart Waste with Junking Plan," *AI*, 23 August 1930, 253–55, 263; "Cleveland Dealers Join Scrap Firm in Junking," *AI*, 26 April 1930, 672 (Cleveland's arrangement involved an independent scrapper, the A. Shaw Company). See also "Would Have Automobile Producers Buy in and Junk Worthless Used Cars," *MA*, 29 January 1915, 36; "Syracuse Dealers to Join Junkers," *MA*, 24 November 1927, 16; "Lansing Salvage Expands," *AI*, 2 June 1928, 854; "La Crosse to Scrap Cars," *AI*, 31 March 1928, 531; and "Cincinnati Studies Car Salvage Plants," *AI*, 10 November 1928, 681.

47. "'Frisco Dealers Burn Old Cars," *MA*, 5 July 1928, 11; Peat, "Dealers Thwart Waste."

48. "$300 Daily Take from Parts Sales," *MA*, 26 January 1928, 17; "St. Louis Yard Buys 2650," *AI*, 12 January 1929, 74; "Omaha Junking Plan"; Zimring, *Cash for Your Trash*, 83 (on Depression-era scrap-metal prices); John B. Rae, *The American Automobile: A Brief History* (Chicago, 1965), 109 (on Depression-era sales and registrations); "2,350,000 Motor Vehicles Scrapped in 1933" (this was down from 2,611,000 in 1932 and 2,927,000 in 1931).

49. This was especially true of the United Auto Wrecking Company of Kansas City. "United Wrecking Co. Called Leader," *MA*, 15 March 1928, 15, 22.

50. "Chevrolet Extending Car Junking Plan," *MA*, 25 August 1927, 20 (quote); "Car Scrapping Enters Inner Circles," *AI*, 20 April 1929, 621.

51. "How the Ford Company Is Helping to Solve 'Used Car Problem,'" *FD&SF*, August 1930, 31–32; "Ford Salvages over 30,000 Used Cars," *AI*, 25 October 1930, 636 (quote); McCarthy, "Henry Ford, Industrial Conservationist."

52. McCarthy, "Henry Ford, Industrial Conservationist," 315.

53. "Seek Idle Metal," *BW*, 31 January 1942, 18; "Getting in the Scrap," *BW*, 14 February 1942, 20; "Scrap Scramble," *BW*, 14 February 1942, 17–18, 20; "Junk Hoard," *BW*, 13 June 1942, 18; Zimring, *Cash for Your Trash*, chap. 4. For a retrospective, see John R. White, "*Hemmings* Offers Seed Money to Help Keep the Old-Car Hobby in Bloom," *Boston Globe*, 4 September 1993.

54. Thomas E. Mullaney, "U.S. to Step Up Auto Graveyard Scrap Flow," *NYT*, 4 November 1951; "NPA to Order More Old Cars Scrapped," *AI*, 15 December 1951, 20; Leonard Westrate, "Scrap Drive Has Passed Peak in Automobile Industry," *AI*, 1 March 1952, 30–31, 60.

55. Average-age and total-registration figures are from "U.S. Motor Vehicle Registrations, 1898–1948" (table), *AI*, 15 March 1949, 91; "Number of Passenger Cars in Use

as of July Each Year" (table), *AI*, 15 March 1948, 120; "Number and Per Cent of Cars in Use, by Age Groups" (table), *AI*, 15 March 1952, 107; "Number and Per Cent of Cars in Use, by Age Groups" (table), *AI*, 15 March 1956, 107; "Number and Per Cent of Car Registrations, by Age Groups" (table), *AI*, 15 March 1960, 92; "Number and Per Cent of Car Registrations, by Age Groups" (table), *AI*, 15 March 1964, 103; and "Number and Per Cent of Car Registrations, by Age Groups" (table), *AI*, 15 March 1968, 112. Scrappage figures are from Marcus Ainsworth, "Total Registration Soars Due to Low Scrappage," *AI*, 15 October 1947, 43, 68; "Vehicle Scrappage in '56 Reached an All-Time High," *AI*, 15 June 1957, 94; and Zimring, *Cash for Your Trash*, 140.

56. Scrap prices vis-à-vis junkyard inventories in the 1950s and 1960s: McCarthy, *Auto Mania*, 154–55; Zimring, *Cash for Your Trash*, 107.

57. John Christy, "What You Can Find in a Wrecking Yard," *ML*, April 1955, 42–43; Jack Phelps, "Bargains in Horsepower," *ML*, November 1955, 26–27, 66; Len Prokine, "Bargains in Used Parts," *Auto Age*, August 1956, 24–25, 54; Joe Petrovec, "Shop JUNKYARDS and Save," *Auto Age*, February 1957, 20–23 (quote on 20); Alex Walordy, "Replacement Parts for Your Car," *CL*, September 1958, 32–33, 60–63.

58. Prokine, "Bargains in Used Parts," 24–25 (quotes); Petrovec, "Shop JUNKYARDS," 21.

59. Prokine, "Bargains in Used Parts," 54 (first quote); Petrovec, "Shop JUNKYARDS," 21 (second quote); Christy, "What You Can Find in a Wrecking Yard," 43.

60. Petrovec, "Shop JUNKYARDS," 23.

61. Quoted in Christy, "What You Can Find in a Wrecking Yard," 43.

62. Zoning, licensing, and space: Howard Lee McBain, "Law-Making by Property Owners," *Political Science Quarterly* 36 (December 1921): 617–41; "Class Dealers as Junk Men," *MA*, 5 August 1915, 17; "Ways of Controlling Automobile 'Graveyards'"; George S. Wehrwein, "The Rural-Urban Fringe," *Economic Geography* 18 (July 1942): 227; Walter Firey, "Ecological Considerations in Planning for Rurban Fringes," *American Sociological Review* 11 (August 1946): 411–23; Zimring, *Cash for Your Trash*, 69–70, 75; McCarthy, *Auto Mania*, 85–86. Salvage titles: John C. Gourlie, "Resale of Cars Sold as 'Junk' Preventable in Most States," *AI*, 26 May 1928, 796–98.

63. The abandoned-car problem: Philip G. Zimbardo, "A Field Experiment in Auto Shaping," in *Vandalism*, ed. Colin Ward (London, 1973), 85–90; Larry Ford and Ernst Griffin, "The Ghettoization of Paradise," *Geographical Review* 69 (April 1979): 140–58; Kevin Lynch with Michael Southworth, *Wasting Away: An Exploration of Waste* (San Francisco, 1990), 62–63; Zimring, *Cash for Your Trash*, 107; and McCarthy, *Auto Mania*, 86, 153–55. The Highway Beautification Act: Zimring, *Cash for Your Trash*, chap. 5; Zimring, "Neon, Junk, and Ruined Landscape," 94–107; and McCarthy, *Auto Mania*, chap. 8. Shredders: Zimring, "The Complex Environmental Legacy of the Automobile Shredder," *Technology and Culture* 52 (July 2011): 523–47. The removal business: Cartakers advertisement, *C&P*, October 1975, 60.

64. Eileen Keerdoja, "Lady Bird's Bill," *Newsweek*, 5 March 1979, 18; Zimring, "Neon, Junk, and Ruined Landscape," 104–5; Zimring, "Complex Environmental Legacy," 529–30. Some were disappointed in the bill but willing to defend the necessity of salvage businesses. See Ralph Blumenthal, "The Junkyard, Ugly and Visible as Ever, Called Vital," *NYT*, 29 June 1975; and below, chap. 5.

65. Zimring, *Cash for Your Trash*, 148–50.

66. James R. Chiles, "Great American Junkyard," *Smithsonian*, March 1985, esp. 54–55; Randy Moser, "Desert Valley Auto Parts," *C&P*, June 1994, esp. 7; Bob Stevens, "Memory Lane Collector Car Dismantlers," *C&P*, January 1996, esp. 57; Bob Howard, "Inland Empire Focus Everything but the Dog," *Business Press* (Ontario, Calif.), 25 March 1996, 1; Skinner, "Renegade Wrecker," esp. 56–57.

67. A few examples from among the most egregious cases: Mike Copeland, "Tainted Soil Ordered Removed," *SPT*, 28 July 1987; John P. Martin, "Junkyard Is Told to Halt Dumping," *PI*, 30 May 1991; Dan DeWitt, "Junkyard Faces Trouble Again," *SPT*, 31 May 1992; Cyril T. Zaneski, "Pompano Junkyard Hit with a $500,500 Fine," *Miami Herald*, 15 April 1993; "Pennsylvania Cleanup Needs Subcontractors," *Superfund Week*, 20 February 1998; Ryan Huff, "Atascadero, Calif., Junkyard Collects $3.5 Million in Hazardous-Waste Fines," *Tribune* (San Luis Obispo, Calif.), 26 May 2004; and "State Fines Auto Junkyard for Improper Waste System," *Honolulu Advertiser*, 15 December 2009.

68. "Murder Victim's Body Found at Auto Junkyard," *SPT*, 18 December 1987; Robert D. McFadden, "Grand Jury Writs Served on Hundreds of Members of 5 Mafia Families Here," *NYT*, 17 October 1972; James M. Markham, "Police Suspend Lieutenant for Alleged Link to Mafia," *NYT*, 19 October 1972; Markham, "Politicians Cited in Mafia Inquiry," *NYT*, 20 October 1972; Linda Wheeler, "Md. Junkyard Auto Crusher Digests Stolen Car with No Questions Asked," *WP*, 13 April 1982; "122 Stung by 'Stolen' Cars," *NYT*, 23 December 1984; Karen Blumenthal, "Turning Your Car into a Convertible—at No Cost," *WSJ*, 10 August 1987; Junda Woo and Dave Kansas, "Law," *WSJ*, 3 April 1992; Norimitsu Onishi, "Your Car, the Sitting Duck," *NYT*, 19 March 1995. See also John A. Heitmann and Rebecca H. Morales, *Stealing Cars: Technology and Society from the Model T to the Gran Torino* (Baltimore, 2014), esp. chap. 4.

69. Newton Auto Salvage Company advertisements, *MA*, 26 June and 25 December 1919. See also Import Auto Parts classifieds, *C&D*, 1975–1978.

70. Louise Saul, "Treasure in an Auto Junkyard," *NYT*, 5 December 1976. See also All Auto Parts Company classified, *R&T*, May 1977, 161.

71. Quoted in Paul Hemp, "Junkyards Polish Their Rusty Image," *NYT*, 30 August 1983.

72. Fred McMorrow, "Survival Is Tough at Auto Graveyards," *NYT*, 1 February 1981; Howard Scott, "Fleetline Automotive," *C&P*, September 1990, esp. 14; Dean Shipley, "A Lotta Auto Parts," *C&P*, December 1997, 40–43.

73. Hemp, "Junkyards Polish Their Rusty Image."

74. David Wessel, "Computer Finds a Role in Buying and Selling, Reshaping Businesses," *WSJ*, 18 March 1987; Peter Rolph, "Dismantlers Clear the Roads," *AN*, 19 November 1989, 235; Jeff Babcock, "Revolvstore," *C&P*, April 1996, 15–17; Haynes, "Yesterday's Junkyards Are Today's High-Tech Recyclers."

75. Dan Abood, "Gold's Auto Wrecking," *C&P*, November 1999, 60–61, 63; United Auto Recyclers Group classifieds, *C&D* and *R&T*, 1997–1998; Hotchkiss, "Buy Auto Parts, New, Used over Internet."

76. Kevin Ross, "Vogt's Parts Barn," *C&P*, March 1990, 6–8, 10; Nugent, "Auto Recycler Defies Stereotype"; Bob Stevens, "American Parts Depot," *C&P*, September 1994, 10; Dean Shipley, "East West Auto Parts," *C&P*, November 1995, 8–10, 12.

77. Robert Louis Adams, "An Economic Analysis of the Junk Automobile Problem" (Ph.D. diss., University of Illinois at Urbana-Champaign, 1972), 3–4; Bob Stevens, "Visiting the 'Great Wall,'" *C&P*, May 1991, esp. 40; Bob Howard, "Junkyards Going High-Tech," *Business Press*, 25 March 1996; Bob Stevens, "A Real Graveyard for Vintage Cars," *C&P*, August 1998, esp. 49.

78. Bob Stevens, "Coast G.M. Salvage," *C&P*, September 1991, 62–65; Skinner, "Renegade Wrecker"; Bob Stevens, "Papke Enterprises Might Just Be the World's Smallest Salvage Yard," *C&P*, April 1991, 58–60, 62.

79. Stevens, "Coast G.M."; Lynne K. Varner, "Junkyard Stirs Ire of Neighbors," *WP*, 9 April 1992; Karen A. Holness, "DEEDCO Fights Plan to Build a Junkyard," *Miami Times*, 19 June 1997; Dana Hedgpeth, "Knowing What's Hot Is the Key to Auto Recycling," *WP*, 28 August 2006.

80. Richard Sisson, "Maryland Yard Located," *C&P*, July 1990, 170; West Peterson, "Leo Winakor & Sons, Inc.," *C&P*, April 2000, 52–53, 55; Joe Sharretts, "Smith Brothers Used Auto Parts," *C&P*, October 2007, 68–72; Ron Kowalke, "Kansas Yard Needs Your Help," *OCW*, 6 May 2010, 20–22.

81. Hof, "Turning Rust into Gold"; "Copart to Celebrate 25 Years of Success," *Collision Repair Magazine*, 17 July 2007, http://collisionrepairmag.com/component/content/article/66-profiles/10359-copart-to-celebrate-25-years-of-success (accessed 9 January 2015); George P. Blumberg, "Junkyards Discard an Image, and the Scary Dogs, Too," *NYT*, 23 October 2002; Hunsberger, "Profits in Parts"; Joseph A. Slobodzian, "Junkyard Proving a Hard Sell in Phila.," *PI*, 31 August 2006.

82. Bob Stevens, At the Wheel, *C&P*, August 1989, 69 (quotes). See also Joe Mayall, Curbside, *SS*, May 1981, 6; Ross, "Vogt's Parts Barn"; and Peterson, "Leo Winakor & Sons." *Cars and Parts* ran features on hundreds of salvage yards in the 1980s, 1990s, and 2000s, publishing at least one in nearly every issue over those decades; see below, chaps. 3–4.

83. Stevens, At the Wheel, 69.

2. Parts, Parts Cars, and Car Enthusiasts

1. Gregory Crouch, "Deep Junkyard," *International Herald Tribune*, 9 August 2003; "Okay, Sure, There May Be Some Water Damage," *C&D*, January 2004, 69; John Pearley Huffman, "Zoom-Zoom-Zoom to Doom," *C&D*, October 2008, 126–30.

2. Life expectancies have varied over the years, from just under 6 in 1958 ("Shorties," *AI*, 1 July 1958, 85) to 9.1 in 2002 (Les Jackson, "Take Notice of Another 'Scrap Bill,'" *WT*, 8 February 2002).

3. On the phenomenon of the "barn find," see below, chap. 4.

4. James R. Chiles, "Great American Junkyard," *Smithsonian*, March 1985, 60–61.

5. See, for example, Robert Jay Stevens, "Easy Jack Antique Auto Parts," *C&P*, February 1982, 50–54; Bob Stevens, "Wilhelm's Wrecking," *C&P*, August 1990, 16; and Lyle R. Rolfe, "Ace Auto Salvage," *C&P*, March 2008, 67.

6. On the scrap-metal trade, see Carl Zimring, *Cash for Your Trash: Scrap Recycling in America* (New Brunswick, N.J., 2005); and Zimring, "The Complex Environmental Legacy of the Automobile Shredder," *Technology and Culture* 52 (July 2011): 523–47.

7. Dan Fisher, "There May Be a Lot of Recycled Detroit in That Japanese Import," *LAT*, 23 April 1972.

8. Zimring, *Cash for Your Trash*; Zimring, "Complex Environmental Legacy."

9. Verni Greenfield, *Making Do or Making Art? A Study of American Recycling* (Ann Arbor, [1984] 1986); Charlene Cerny and Suzanne Seriff, eds., *Recycled, Re-seen: Folk Art from the Global Scrap Heap* (New York, 1996).

10. Dolly Jørgensen, "An Oasis in a Watery Desert? Discourses on an Industrial Ecosystem in the Gulf of Mexico Rigs-to-Reefs Program," *History and Technology* 25 (December 2009): 343–64; "10 Best Frozen, Lonely, and Mongrelly Merged Cars," *C&D*, January 2007, 69.

11. Chiles, "Great American Junkyard," 62; Todd Howard, "*Hot Rod*'s Route 66 Tour, Part 3," *HRM*, February 1988, esp. 59, 65; Bob Stevens, "Touring Route 66, Part 7," *C&P*, February 2001, 58–60; Rich Ceppos, "Ten Best Cartifacts," *C&D*, January 1989, 64; Stephen Kim, Roddin' at Random, *HRM*, June 2008, 38 (quotes).

12. Philip Quinn, "Carhenge," *National Post*, 18 January 2002; News and Notes, *C&P*, December 2005, 14; Phil Berg, "Carchitectural Wonders," *C&D*, January 1991, 57–60; PS, *R&T*, April 1981, 184; Reader Sightings, *C&D*, September 1992, 12; Jay J. Hector, "Long Term Parking," *R&T*, August 1984, 58; Ceppos, "Ten Best Cartifacts," 65.

13. Andrew Bornhop, "Parts," *R&T*, March 2011, 71; Henry Rasmusen and Art Grant, *Sculpture from Junk* (New York, 1967), 6–7; Mary Lou Stribling, *Art from Found Materials: Discarded and Natural* (New York, 1970); Lea Vergine, ed., *Trash: From Junk to Art* (Corte Madera, Calif., 1997); "10 Best Things Made from Car Parts," *C&D*, January 2002, 60–61; Mick Walsh, "Non Sequitur," *Automobile*, November 2005, 102; John Kearney, Sculptor, www.johnkearneysculptor.com (accessed 19 January 2015).

14. "Rustart," *C&D*, March 1973, 74–77 (quotes on 74).

15. John Pearley Huffman, "Midnight in the Garden of Eldorados and E-Types," *C&D*, September 2005, 112–16, 118 (quotes on 115 and 118). Similar collections were assembled elsewhere—on the other side of the world, for example, by a man in Kraaifontein, South Africa, who "prefers old cars to rust in peace than go to the crusher" ("The Cape's Old-Car Crusader," *C&SC*, February 2011, 51). Closer to home, gearheads in South Carolina, Illinois, California, Oregon, and elsewhere did the same. See Randy Moser, "Home of Many Rare Treasures," *C&P*, November 1993, 52–54, 56, 58; J. James Fontana, "Detroit Orphanage," *C&P*, May 1990, 56–58, 60; Joseph J. Caro, "Welcome to Glen's," *C&P*, May 1990, 28–30; Mark J. Hash, "Willie's Salvage and 'Museum,'" *C&P*, November 1991, 36–38, 40, 42; and below, chap. 4.

16. Stevens, "Easy Jack Antique Auto Parts," 50.

17. "What Becomes of Old Automobiles," *MA*, 23 February 1928, 12; "Old Chassis Sawing Wood," *HA*, 24 September 1913, 509; "Takes Model T's Off the Road, Puts Them to Work for Farmers," *FD&SF*, December 1930, 54; "No Rest for an Old Friend," *FD&SF*, March 1932, 28; "Once More the Model T," *FD&SF*, November 1931, 44; "A Department of Better Business," *MA*, 5 May 1921, 34; "New Use for Old Car," *TA*, 3 August 1905, 143; "Transforming Used Passenger Cars into Trucks," *HA*, 1 March 1918, 29–31; "Something Old, Something New," *R&C*, October 1953, 46–49.

18. PS, *R&T*, October 1996, 178; PS, *R&T*, August 1987, 168; Sarah A. Webster, "Car Parts Artist," *Detroit Free Press*, 4 February 2010; Chiles, "Great American Junkyard,"

62; Linda Barnard, "It Started with a Citroën," *Toronto Star*, 15 March 2003; Reader Sightings, *C&D*, December 1994, 14; Berg, "Carchitectural Wonders," 58.

19. Advertisements: Terry Shuler, Eastern Scene, *HVW*, August 1991, 15; Joe Pidd, Bug Mail, *HVW*, November 1991, 8; Chris Hersh, Bug Mail, *HVW*, December 1993, 8; David Heard, Transporter Talk, *HVW*, November 1994, 34; Fertig, *VWT*, May 1996, 109. Gadzooks: Tiffany Alonzo, Bug Mail, *HVW*, June 1992, 8; Fertig, *VWT*, July 1997, 111. Bead-blasting hood: Rich Kimball, Collector's Corner, *HVW*, June 1994, 32. Chicken coops: PS, *R&T*, December 1983, 204; Tom Medley, "Chicken Coop Tudor," *R&C*, 1977, 20–21 (at this point, *Rod and Custom* appeared annually); Gray Baskerville, "The Fordgatherers," *HRM*, August 2000, 64. Wheelbarrow: K. Taylor, Bug Mail, *HVW*, July 1996, 13. Tractor shade: PS, *R&T*, July 1987, 176. Motor homes: Fertig, *VWT*, December 1994, 107; Fertig, *VWT*, November 1996, 109. Trailers: Charlie Robinton, Bug Mail, *HVW*, May 1996, 11; Ed Kreb, "What Not to Do with Your VW," *VWT*, July 1993, 26. Treehouse: Terry Shuler, Eastern Scene, *HVW*, December 1993, 40.

20. Tom McCarthy, *Auto Mania: Cars, Consumers, and the Environment* (New Haven, Conn., 2007), 155.

21. Hot rodding was and would remain a pastime dominated by men, as would the other enthusiast niches described in this chapter: customization, restoration, sports cars and imports, and street rods. To be sure, women have long been active in organized motor sports, from Veda Orr on the dry lakes of Southern California in the 1930s and 1940s to Denise McCluggage on sports-car circuits in the 1950s; Shirley Muldowney on the drag strips in the 1960s, 1970s, and 1980s; and Danica Patrick on IndyCar and NASCAR circuits beginning in the 2000s, among countless others. Nevertheless, whether in the workshop, at a car show, in the pits, or on the track, the various activities associated with automobile enthusiasm have long been dominated by masculine energy—the gearhead universe, in other words, is largely the preserve of "boys and their toys." See H. F. Moorhouse, "Racing for a Sign: Defining the 'Hot Rod,' 1945–1960," *Journal of Popular Culture* 20 (1986): 83–96; Moorhouse, "The 'Work' Ethic and 'Leisure' Activity: The Hot Rod in Post-War America," in *The Historical Meanings of Work*, ed. Patrick Joyce (New York, 1987), 237–57; Moorhouse, *Driving Ambitions: A Social Analysis of the American Hot Rod Enthusiasm* (New York, 1991); David N. Lucsko, *The Business of Speed: The Hot Rod Industry in America, 1915–1990* (Baltimore, 2008), 59; Robert C. Post, *High Performance: The Culture and Technology of Drag Racing, 1950–1990* (Baltimore, 1994), chap. 12; Ben A. Shackleford, "Masculinity, the Auto Racing Fraternity, and the Technological Sublime: The Pit Stop as a Celebration of Social Roles," in *Boys and Their Toys? Masculinity, Class, and Technology in America*, ed. Roger Horowitz (New York, 2001), 229–50; Todd McCarthy, *Fast Women: The Legendary Ladies of Racing* (New York, 2007); and Martha Kreszock, Suzanne Wise, and Margaret Freeman, "Just a Good Ol' Gal: Pioneer Racer Louise Smith," 105–23, John Edwin Mason, "'Anything but a Novelty': Women, Girls, and Friday Night Drag Racing," 125–31, and Patricia Lee Yongue, "'Way Tight' or 'Wicked Loose'? Reading NASCAR's Masculinities," 133–46, all in *Motorsports and American Culture: From Demolition Derbies to NASCAR*, ed. Mark D. Howell and John D. Miller (Lanham, Md., 2014).

22. Lucsko, *Business of Speed*, chaps. 1–3; Post, *High Performance*, chaps 1–3 and p. 126.

23. Lucsko, *Business of Speed*, chaps. 1–2.

24. Ibid.

25. See, for example, Art Bagnall, *Roy Richter: Striving for Excellence* (Los Alamitos, Calif., 1990), 8–9; and Philip Linhares, "Hot Rods and Customs: From the Garage to the Museum," in *Hot Rods and Customs: The Men and Machines of California's Car Culture*, ed. Michael Dobrin, Philip E. Linhares, and Pat Ganahl (Oakland, 1996), 14.

26. On hot rodding's postwar niches, see Lucsko, *Business of Speed*, chaps. 3–10; Don Montgomery, *Hot Rod Memories: Relived Again* (Fallbrook, Calif., 1991), 13–14; Post, *High Performance*, chaps. 1–3; Jessie Embry, "The Last Amateur Sport: Automobile Racing on the Bonneville Salt Flats," *Americana: The Journal of American Popular Culture* 2 (2003), www.americanpopularculture.com/journal.articles/fall_2003/embry.htm (accessed 12 January 2006); John De Witt, *Cool Cars, High Art: The Rise of Kustom Kulture* (Jackson, Miss., 2002); and Andy Southard and Dain Gingerelli, *The Oakland Roadster Show: 50 Years of Hot Rods and Customs* (Osceola, Wis., 1998).

27. Lucsko, *Business of Speed*, chap. 2.

28. Alan L. Colvin, "The Phantom Convertible," *CHP*, December 1999, 109, 111–13; Alex Rogan, "Smoke Signals," *Eurotuner*, February 2008, 36–40; Colin Ryan, "Raw Material," *Import Tuner*, February 2010, 34; "In the Garage," *Eurotuner*, March 2007, 24; O. Dewayne Davis, "Virginia's Ferncliff Auto Parts," *C&P*, September 2004, 50–53; Douglas Kott, Ampersand, *R&T*, November 1998, 37.

29. Victor Max, "Choosing a 914," *EC*, October 1991, 66–68; "914 Tech Tips," *EC*, October 1991, 69–70, 126–29; Victor Max, "One Pair, Six High," *EC*, February 1992, 34–36, 101–2.

30. Rob Kinnan, "Crusher Camaro, Part 3," *HRM*, February 1995, 46–48; Bruce Caldwell, "The Mustang Market," *HRM*, December 1978, 25–28. See also Marlan Davis, "Meaner Than a Junkyard Dog," *HRM*, November 1984, 45–48, 50; and Davis, "10 Most Overlooked Street Machines," *HRM*, December 1985, 28–30, 32–33, 37–40, 42–44.

31. Jim Losee, "Bolt-On Brake Improvements," *PHR*, November 1984, 74–76, 103; Losee, "Identifying High-Performance Rear Ends," *PHR*, February 1984, 46–48; Cam Benty, "The Salvage Yard Maze," *PHR*, August 1982, 76–79.

32. Don Francisco, "Engine Installation," *HRM*, April 1953, 18–21, 67–70, 78 (quote on 18). See also "Fordebaker," *Hop Up*, May 1953, 34–35.

33. Joe Moore, "Bu-Merc Conversion," *ML*, January 1954, 40–41, 54; Reader's Car of the Month, *R&C*, March 1955, 24–25; "One T-Bone, RARE . . . but Well Done," *R&C*, November 1954, 22–25; "Fire Powering the Deuce," *R&C*, October 1954, 46–49; Eric Rickman, "Build a Windsor-Merc," *HRM*, December 1956, 22–25.

34. Lucsko, *Business of Speed*, chap. 5; Ken Gross, "The Day the Flatheads Died," *Hemmings Muscle Machines*, December 2003, 45; Dean Brown, "The Best Engines for Hot Rods," *PHR*, August 1962, 10–15, 70; "A Street Rod for under 10K," *HRM*, October 1994, 31; JCI advertisements, *Hemmings*, 2003–2009; James Sly, "Getting Technical," *EC*, June 1997, 4, 8, 11; Marlan Davis, "Aftermarket Parts Make Swaps Easy," *HRM*, May 1985, 31; Sherri Collins, "Project Y2K 914-8," *EC*, December 1999, 32; John Thawley, "It's a Chevrostang!" *HRM*, September 1969, 92–93; "Beetle Bomb," *HRM*, July 1985, 82; Dean Kirsten, "Adding Sting to the Bug," *C&SC*, August 2013, 166–67, 169, 171; "Range Pan-

zer," *EC*, December 1991, 28–31; Jeff Hartman, "Lotus Europa Engine Swap," *EC*, February 1995, 36–38; Bob Leif, "Dolling Up the Bug," *HRIN*, December 1966, 22–27, 48, 52–53.

35. Don Francisco, "How to Buy Used Engines," *CC*, February 1957, 12–17, 58–59; Ask Bill Frick, *Cars*, January 1960, 4–5; Davis, "Aftermarket Parts Make Swaps Easy."

36. Tex Smith, "A Guide to Best Buys in . . . USED PARTS," *PHR*, August 1967, 63.

37. Kevin Clemens, On the Line, *EC*, June 2001, 6, 8; "Hammer Rabbit V8," *Eurotuner*, June 2007, 20.

38. On customs, see DeWitt, *Cool Cars, High Art*; and Brenda Jo Bright, *Customized: Art Inspired by Hot Rods, Low Riders, and American Car Culture* (New York, 2000).

39. Bud Unger, "Customize It Yourself," *Speed Age*, May 1953, 70–72.

40. Peter Sukalac, "Holiday Theme," *HRM*, May 1955, 26–27.

41. "The Long, Low Look," *HRM*, September 1951, 13.

42. W. O. Boyles, "Lincoln-Ford Hybrid," *R&C*, August 1953, 50–51 (quotes on 50).

43. Robert R. Hovey, "Economy Custom," *Auto Age*, June 1953, 6.

44. Tom Medley, "Practical Pick-Up," *HRM*, February 1950, 17.

45. Orin Tramz, "10 Steps to Customizing," *R&C*, October 1953, 28–33 (quotes on 32 and 33).

46. Felix Zelenka, "Torch Tips," *CC*, December 1953, 24–27; "1942 Buick," *Hop Up*, November 1951, 38.

47. By "real-world bricolage," I mean the fabrication of creative technical solutions from cast-off parts originally designed for other purposes. See Douglas Harper, *Working Knowledge: Skill and Community in a Small Shop* (Berkeley, Calif., 1987); and Susan Strasser, *Waste and Want: A Social History of Trash* (New York, 1999), 22–23; compare with Lisa Parks, "Falling Apart: Electronics Salvaging and the Global Media Economy," in *Residual Media*, ed. Charles R. Acland (Minneapolis, 2007), 43, where she discusses the entertainment-oriented bricolage featured on television's *Junkyard Wars*.

48. Todd Kaho, "Short 'n Sweet," *HVW*, October 1992, 90–91, 112; Rich Van Crafton, "Prime Rare," *VWT*, April 1991, 88–89; Robert K. Smith, "Rocky Mountain Roadsters," *HVW*, March 1993, 72–77; "Reincarnation," *PHR*, May 1976, 54–56; Early Iron, *Street Rodder*, January 1986, 98–103.

49. Rich Kimball, "Just a Daily Driver," *HVW*, October 1990, 74–75, 110 (quotes on 75).

50. Robert K. Smith, "Dream-Come-True Oval Window," *HVW*, August 1994, 46–49; Dean Kirsten, "Litchfield Bug Inn," *HVW*, October 1994, 88–91; Rich Van Crafton, "Der Meter's Running, Bub," *VWT*, February 1991, 48–49, 93.

51. "Double-Deluxe," *HVW*, March 2002, 80–83, 108–9.

52. Dean Kirsten, "'63 Deluxe Crewcab," *HVW*, August 1996, 62–64, 102.

53. Edwin Kilburn, "Fitting a Mechanical Lubricator to an Old Car," *HA*, 20 March 1907, 399–400; C. Teachont, The Reader's Clearing House, *MA*, 4 November 1915, 38; Editorial Perspectives, *MA*, 7 March 1918, 10.

54. "Remaking a 1910 Model," *MA*, 31 May 1917, 47; Ray A. Barnes, Comments and Queries, *HA*, 28 January 1914, 164–65; "Putting the Used Car among Dealer's Assets," *HA*, 1 November 1917, 18–19, 88. See also above, chap. 1, and Steven M. Gelber, *Horse Trading in the Age of Cars: Men in the Marketplace* (Baltimore, 2008), chap. 4.

55. "Old Autocar Recalls Ancient History," *TA*, 16 April 1908, 552; "Motor Car Shop Kinks," *MA*, 8 April 1909, 38; "Where the Old Cars Go," *TA*, 20 May 1909, 841.

56. "What Becomes of Out of Date Cars," *HA*, 31 May 1905, 600 (quote); "What Becomes of Old Automobiles"; "Where the Old Cars Go."

57. The earliest coverage of restoration as such (as opposed to refurbishing) that I found appeared in "Motor Junk Is Sold at Fair," *MA*, 3 June 1915, 28.

58. "England Opens Its Motor Car Museum," *MA*, 27 June 1912, 22 (quote); W. S. Fowler Dixon, "John Bull Preserves Early Models of Power Propelled Vehicles by Placing Them in Motor Museum," *MA*, 2 April 1914, 20–21.

59. "Pennsylvania's 'Oldest Car,'" *HA*, 28 January 1914, 148.

60. From the Four Winds, *MA*, 8 June 1916, 47; "New Haynes for Old," *MA*, 19 October 1916, 24.

61. "Ancient History," *TA*, 23 February 1911, 539, and 2 March 1911, 627; "Motoring of Early Days," *MA*, 6 June 1918, 5–11. See also Charles E. Duryea, "Motor Racing 20 Years Ago," *MA*, 27 May 1915, 24–28.

62. "The Oldest Car in Service," *Motor*, May 1911, 53; "Priest Owns Oldest Car in the World," *MA*, 27 June 1912, 22–23; "Thirty Years' Constant Service from One Car," *AI–TA*, 7 July 1921, 33; "10-Year-Old Cadillac Makes Fine Showing in English Test," *TA*, 6 November 1913, 883; "A Venerable Ride," *Motor*, July 1913, 79–80; From the Four Winds, *MA*, 12 August 1920, 56.

63. Bob Stubenrauch, *The Fun of Old Cars: Collecting and Restoring Antique, Classic and Special Interest Automobiles* (New York, 1967), 33. These clubs deserve their own scholarly treatment, but to date they have barely made it into the literature at all apart from fragmentary periodical evidence, the clubs' own self-portraits (www.aaca.org, www.hcca.org), and enthusiast-oriented books like Stubenrauch's.

64. These period definitions are based on Eugene Jaderquist, "The Classic Thrill of Yesteryear," *True's Automotive Yearbook*, 1952, 2–7, 100–102; "How to Restore a $50.00 Classic," *MT*, February 1953, 54–57; Joe H. Wherry, "Classic Cars from the Vintage Years," *CL*, September 1954, 52–55; Robert H. Gottlieb, "So Who Wants an Old Car?" *MT*, November 1957, 42–43; Menno Duerksen, "Free Wheeling," *C&P*, May 1969, 45–46; Carl Cramer and Menno Duerksen, Tool Bag, *C&P*, October 1969, 45; and Switzer and Duerksen, Tool Bag, *C&P*, September 1971, 68. See also John B. Rae, *The American Automobile: A Brief History* (Chicago, 1965), 217–18. By the 1970s and 1980s, "classic car" had assumed a much broader meaning; see below, chap. 6.

65. Witness the advent of a *Hemmings* spinoff in 1970, *Special-Interest Autos*, which focused on the mass-market makes of the 1920s–1950s.

66. Jaderquist, "Classic Thrill of Yesteryear," 100–102 (quote on 100). Median-income data are from U.S. Department of Commerce, "Family Income in the United States: 1952," press release, 27 April 1954, available online at www2.census.gov/prod2/popscan/p60-015.pdf (accessed 26 January 2015).

67. "Restoring an Aristocrat," *ML*, January 1956, 50–51; "Dealer or Dabbler," *Motorsport*, June–July 1955, 28–30.

68. Frank Cetin, "He Brings 'Em Back to Life," *Cars*, June 1953, 53.

69. "The Past Meets the Present," *R&C*, February 1954, 10.

70. On manufacturer- and dealer-led scrap-and-salvage operations in the 1920s and 1930s, see above, chap. 1, and Gelber, *Horse Trading*, 71–72. On World War II, see John R. White, "*Hemmings* Offers Seed Money to Help Keep the Old-Car Hobby in Bloom," *Boston Globe*, 4 September 1993; and Zimring, *Cash for Your Trash*, chap. 4. On the Korean War, see Thomas E. Mullaney, "U.S. to Step Up Auto-Graveyard Scrap Flow," *NYT*, 4 November 1951; "NPA to Order More Old Cars Scrapped," *AI*, 15 December 1951, 20; and Leonard Westrate, "Scrap Drive Has Passed Peak in Automobile Industry," *AI*, 1 March 1952, 30–31, 60. For a contemporary perspective on the Korean War, see Robert J. Gottlieb, Classic Comments, *MT*, October 1953, 36–37, 55.

71. "The Past Meets the Present," 10; Charles L. Stratton, "Museum in the Rough," *Motorsport*, August 1953, 49.

72. George A. Parks, "Perpetual Youngsters," *CL*, December 1954, 42–45 (quote on 43).

73. Joe H. Wherry, "Model A Club," *CL*, August 1954, 51–54 (quote on 52); Gene Jaderquist, "Classics and Antiques," *Motorsport*, November–December 1954, 33–36 (quote on 36). See also "Long Live the A," *Auto Craftsman*, December 1957, 12–13.

74. Peter Bohr, "Factory Restoration," *R&T*, May 2003, 121; Jim Mateja, "GM Does Its Part for Restoring Older Vehicles," *CT*, 27 September 1996; Ford Motor Company, "Restoration Parts Licensing Program," www.fordrestorationparts.com/restoration/Restoration_Parts_Licensing_Program.pdf (accessed 18 May 2015); MK1 Autohaus, "Announcing Our Partnership with VW Classic Parts," www.mk1autohaus.com/VW-Classic-Parts_ep_46.html (accessed 18 May 2015); Ferrari.com, "More Than 60 Years of Memorable Models," http://auto.ferrari.com/en_EN/sports-cars-models/past-models/ (accessed 18 May 2015).

75. Caldwell, "Mustang Market"; Tom Senter, "Buying a Bent Car," *PHR*, December 1980, 63–65, 79–80, 82–83. On *Cars and Parts* and enthusiast-oriented salvage yards, see below, chap. 3.

76. This section draws heavily from my essay "American Motor Sport: The Checkered Literature on the Checkered Flag," in *A Companion to American Sport History*, ed. Steven A. Riess (Chichester, U.K., 2014), 326–27. See also Ken W. Purdy, *The Kings of the Road* (Boston, 1949); William F. Nolan, *Men of Thunder: Fabled Daredevils of Motor Sport* (New York, 1964); Brock Yates, *Against Death and Time: One Fatal Season in Racing's Glory Years* (New York, 2004); Todd McCarthy, *Fast Women: The Legendary Ladies of Racing* (New York, 2007); and especially Jeremy Kinney, "Racing on Runways: The Strategic Air Command and Sports Car Racing in the 1950s," *ICON* 19 (2013): 193–215. *Sports Cars Illustrated* became *Car and Driver* in April 1961.

77. See, for example, Janius G. Eyerman, Tool Bag, *C&P*, June 1970, 46.

78. Henry Z. De Kuyper, "Micro Magic," *VWT*, April 1991, 66–67, 98; Jay Jones, "2002 Transformed," *EC*, July–August 1991, 76–77, 108–9; Brett Johnson, "1951 356 Porsche, Part 1," *EC*, September 1993, 66–68.

79. "A 'D' for Devin," *MT*, April 1960, 81; "Quickie Desert Wagon," *HRM*, April 1964, 70–73, 96; Don Emmons, "Let's Build an Ultra-Light Superbug," *R&C*, April 1968, 58–61, 74; John McElroy, "Individuality at a Price," *AI*, November 1981, 55–57; Michael Jordan, "Reincarnated Cars," *C&D*, March 1982, 79–82, 84–85.

80. McElroy, "Individuality at a Price"; Jordan, "Reincarnated Cars."

81. "A 'D' for Devin," 81.

82. "Quickie Desert Wagon."

83. Bruce Simurda, "Pocket Rockets," *HVW*, 48–51, 118, 124–25; Myers Manx, "Our History," www.meyersmanx.com/history.shtml (accessed 27 January 2015); Jordan, "Reincarnated Cars"; Bradley GT advertisement, *C&P*, December 1975, 13.

84. Hot rodding's postwar niches: see above, chap. 2 note 26. Detroit's muscle cars: Roger Huntington, *American Supercar: Development of the Detroit High-Performance Car* (Osceola, Wis., 1983). Hot rodders' reactions to muscle cars: Ron Roberson, *Middletown Pacemakers: The Story of an Ohio Hot Rod Club* (Chicago, 2002), chap. 8; Lucsko, *Business of Speed*, chaps. 5–6. On the demographics of street rodding in its early years, see Cec Draney, "*Hot Rod Industry News* Survey," *HRIN*, September 1967, 39, according to which the majority (50.8 percent) of street rodders in 1967 were between the ages of eighteen and twenty-four.

85. Joe Mayall, Curbside, *SS*, May 1981, 6.

86. Jerry Hegge, "Jerry Dawson," *R&C*, April 1973, 25–26 (quotes on 25).

87. Gray Baskerville, "Baskerville's Gallery," *HRM*, May 1976, 120.

88. "Big Blue Booth," *HRM*, April 1975, 106–7.

89. Whereas a slight majority (50.8 percent) of street rodders in 1967 were under the age of twenty-five, by 1977 their average age had crept up to thirty-two, and by 1987 a whopping 80 percent were over the age of thirty, including 36 percent over forty. See Draney, "*Hot Rod Industry News* Survey"; "Hot Rod Magazine Street Rod Market Survey," *HRIN*, June 1977, 21; and Joe Mayall, Curbside, *SS*, May 1987, 4.

90. Lucsko, *Business of Speed*, 224–29.

91. Kathleen Piasecki, "The Last Hurrah," *SS*, January 1981, 6.

92. Will O'Neil, "Slow Lane," *SS*, May 1981, 10.

93. Lucsko, *Business of Speed*, 225–26.

94. "The Great Rat-Rod Revival," *HRM*, December 1998, 140, 142, 144–47; David Freiburger, "Order Rodentia," *HRM*, November 2010, 42–48, 50, 52, 54, 56–57; Alex "Axle" Izardi, "They Invented Rat Rodding," *HRM*, April 2011, 19; Michael D. Girardi, "More Derelict," B. Craft, "Just Decay," and Casey Dolin, "Really a Beater," all in *HRM*, June 2011, 20.

95. Bob Chuvarsky, "Buying Protection," *VWT*, December 1997, 56–58; Ray Thursby, "Protecting Your Investment," *EC*, May 1999, 88–91; Ryan Lee Price, "The Black Arts," *EC*, October 2005, 74–76, 78–82. Section title is from Condon and Skelly advertisement, *C&P*, December 1994, 31.

96. Bob Zeller, "A Flood of Fraud," *C&D*, January 2006, 104–8; Chris Woodyard, "Officials on the Trail of Junkers," *USA Today*, 29 May 2008; Angie Favot, "Junker Keeper," *Autoweek*, 19 October 2009, 13. Cars totaled in the United States sometimes end up back in use abroad; see Jim Travers, "Damaged Goods," *MPH*, August 2005, 26–27.

97. Chuvarsky, "Buying Protection," 57.

98. James A. Grundy advertisements, *C&P*, October 1978, 53 (quote) and August 1989, 31.

99. J. C. Taylor advertisement, *C&P*, January 2008, 57.

100. Dempsey and Siders advertisement, *C&P*, November 1993, 17; American Collectors advertisement, *C&P*, March 1996, 67; Condon and Skelly advertisement, *C&P*, May 1996, 59; Thursby, "Protecting Your Investment," 89.

101. Condon and Skelly advertisement, *C&P*, June 1998, 141; Hagerty advertisement, *C&P*, August 1997, 106.

102. Hagerty advertisement, *C&P*, March 1999, 37; Thursby, "Protecting Your Investment," 89.

103. See Harold W. Sauter, untitled open letter, *C&P*, September 1971, 87; "Miscellany," *C&P*, August 1974, 6; Tony Hogg, "Car Speculations," *R&T*, September 1981, 27–28; and Peter Bohr, "Collector Cars," *R&T*, February 1994, 52–61.

104. See, for example, Senter, "Buying a Bent Car"; Dennis Alder, "Fresh Slant," *EC*, April 1992, 28–29, 99; "Quick Recovery," *HRM*, February 1994, 79; and Tim McKinney, "A 1.8T Tale of Addiction," *EC*, September 2004, 60–62.

3. Arizona Gold

1. This boast appeared on the cover of *Hemmings* for many years.

2. Car enthusiasts are fairly evenly distributed among the general population, so the more cars a given area has on the books, the more enthusiasts one is likely to encounter there. Thus there is no reason to expect the volume of cars and parts offered for sale to enthusiasts in a given area to track any higher or lower than that area's share of the national vehicle fleet, yet most states do track higher or lower than a 1:1 ratio. Part of the aim of this chapter is to account for this difference. State registration data: www.fhwa.dot.gov/policyinformation/statistics.cfm (accessed 12 July 2013); deviations from a 1:1 cars-to-classifieds ratio calculated using a differential of 20 percent.

3. Because *Hemmings* is sorted by make, model, and age, many wrecking yards pay for several advertisements each month—one in the Buick section, another in the early Ford section, and so forth.

4. When it comes to automotive rust, snow does not matter as much as one might think. Instead, the two chief culprits are road salt and high humidity, especially coastal humidity.

5. These statistics are derived from an exhaustive survey of advertisements and classifieds from salvage businesses in a broad cross-section of enthusiast periodicals from the 1950s through the early 2010s. Rust Belt, East Coast, and southeastern states that tracked low during this period included Alabama, Arkansas, Connecticut, Hawaii, Illinois, Indiana, Kentucky, Louisiana, Maryland, Massachusetts, Michigan, Mississippi, New Hampshire, New Jersey, New York, North Carolina, Ohio, Pennsylvania, Vermont, and West Virginia; Utah and Texas also tracked low. By contrast, arid states that tracked high included Alaska, Arizona, California, Colorado, Idaho, Kansas, Montana, Nebraska, Nevada, Oklahoma, Oregon, South Dakota, Washington, and Wyoming; Delaware, Georgia, Maine, and Wisconsin also tracked high. Florida, Iowa, Minnesota, Montana, New Mexico, North Carolina, North Dakota, Rhode Island, Tennessee, and Virginia tracked close to 1:1.

6. American Heritage Auto advertisement, *Hemmings*, December 2003, 428; Arizona Mailorder Autoparts advertisement, *C&D*, September 1998, 192; Automania

advertisement, *Hemmings*, November 2007, 306; Arizona Desert Specialties advertisement, *C&P*, December 1990, 77; Fat Albert's advertisement, *C&P*, August 1994, 168.

7. See, for example, Bob Stevens, "Bill Lindsay's Ford Collection," *C&P*, April 1996, 36–38 (Ohio); Murray Park advertisement, *Hemmings*, August 2008, 234 (Ohio); Desert Dog Auto Parts advertisement, *Hemmings*, November 2007, 327 (Illinois); Boos-Herrel advertisement, *Hemmings*, November 2007, 573 (Pennsylvania); and Jim's Auto Center advertisement, *OCW*, 7 May 2009, 28 (Wisconsin).

8. David Edgerton, *The Shock of the Old: Technology and Global History since 1900* (New York, 2007); Michael Thompson, *Rubbish Theory: The Creation and Destruction of Value* (New York, 1979); Kevin Lynch with Michael Southworth, *Wasting Away: An Exploration of Waste* (San Francisco, 1990).

9. Tom McCarthy, *Auto Mania: Cars, Consumers, and the Environment* (New Haven, Conn., 2007); Carl Zimring, *Cash for Your Trash: Scrap Recycling in America* (New Brunswick, N.J., 2005); Zimring, "The Complex Environmental Legacy of the Automobile Shredder," *Technology and Culture* 52 (July 2011): 523–47.

10. Gray's Auto Garage advertisements, *MA*, 1919–1922 (Ford); Fish and Unger Auto Wrecking advertisement, *MA*, 30 December 1920, 129 (Ford); Baird advertisement, *MA*, 23 April 1925, 89 (Franklin).

11. The story of George Wight and his yard is based on "Touring the Hot Rod Shops," *HRM*, April 1951, 32–33; Art Bagnall, *Roy Richter: Striving for Excellence* (Los Alamitos, Calif., 1990), 8–9; Brock Yates, "Reinventing the Wheel," *C&D*, September 1993, 107–10, 112; and David N. Lucsko, *The Business of Speed: The Hot Rod Industry in America, 1915–1990* (Baltimore, 2008), 51–52.

12. Lee Chapel, "Origin of a Speed Shop," *HRM*, June 1928, 13.

13. Sherman Way Auto Wreckers and Grand Prix Auto Parts advertisements and classifieds, *R&T*, *SCI*, and *C&D*, 1956–1971 (quoted material is from the firm's August 1956 spot in *SCI*, 64). See also Russ Kelly and Len Griffing, "The Wrecking Yard Bit," *SCI*, February 1958, 30–31. Grand Prix Auto parts eventually gave way to a Fiat specialist on the same site; see Giant Auto Wreckers advertisements, *R&T* and *C&D*, 1984–1985.

14. Kelly and Griffing, "Wrecking Yard Bit," 30.

15. Jack's Auto Parts classifieds, *SCI*, August and October 1958; All Auto Parts classifieds, *C&D*, 1961–1963; Road and Track Auto Parts classifieds, *C&D*, 1963–1966.

16. Foreign Car Parts Mart classifieds, *SCI* and *C&D*, 1961–1963.

17. A few examples from the scores I have on file: air-cooled VW specialists Strictly Used Volkswagen Parts (classifieds, *HVW*, 1984–1986), Sunray Bugs (advertisements, *VWT*, 1994–1996), and Recycle Auto Parts (advertisements, *HVW*, 1988–1991); water-cooled VW specialists Recycled Rabbit/Wolfsport (advertisements, *VWT*, 1990–1995) and BW Auto Dismantlers (advertisements, *EC*, 1993–2006); Porsche specialists Best Deal, Inc. (classifieds, *R&T*, 1976–1995), Oklahoma Foreign (advertisements, *R&T*, *EC*, and *Excellence*, 1979–2004), Porsche Auto Recyclers (advertisements and classifieds, *C&D* and *R&T*, 1982–1988), Automobile Atlanta (classifieds, *C&D* and *R&T*, 1982–1992), AASE Brothers Porsche Car Dismantlers (advertisements and classifieds, *R&T* and *EC*, 1982–1994), Kempton Brothers Porsche Parts (advertisements, *C&D* and *EC*, 1987–1993), and Porscheaven/Parts Heaven (advertisements, *EC*, 1991–1998); BMW

specialists Bavarian Motor Wrecking (advertisements and classifieds, *EC*, 1992–1996), The Auto Works (advertisements and classifieds, *C&D* and *R&T*, 1994), and Bavarian Auto Recycling (advertisements, *R&T*, *EC*, *Hemmings*, and *C&D*, 1995–2004); Mercedes-Benz specialists Sun Valley Mercedes Dismantlers (classifieds, *R&T* and *C&D*, 1982–1987), Embee Parts (classifieds, *C&D* and *R&T*, 1984–1994), ATVM (classifieds, *C&P*, 1990–1999), and Southern Star (classifieds, *C&P*, 1996–2000); Jaguar specialist Welsh's Jaguar Enterprises (advertisements and classifieds, *C&P*, *C&D*, and *R&T*, 1982–2000); Volvo specialists Voluparts (classifieds, *C&D* and *R&T*, 1981–2000), Hirsch Industries (advertisements and classifieds, *C&D*, *EC*, and *R&T*, 1994–1998), and Revolvstore (Jeff Babcock, "Revolvstore," *C&P*, April 1996, 15–17); Saab specialist Saab Parts (classifieds, *R&T*, 1994–1995); Fiat specialists Asian Italian Auto Parts (advertisements, *C&D* and *R&T*, 1983–1993), Giant Auto Wreckers (classifieds, *C&D* and *R&T*, 1984–1990), and Bayless (classifieds, *R&T*, *C&D*, *EC*, and *Hemmings*, 1990–2007); Mazda specialists Mazdatrix (classified, *R&T*, November 1989, 193); Nissan/Datsun specialists Datsun Z (classifieds, *C&D*, 1988–1998) and Datsun 1600 & 2000 Roadster Parts (classifieds, *C&D* and *R&T*, 1991–1994); British specialists British Auto Salvage (classifieds, *C&P*, 1977), British Foreign Auto Salvage (classifieds, *C&D* and *R&T*, 1979–1983), British Auto Wreckers/British Auto Center (classifieds, *R&T* and *C&D*, 1980–1991), and FASPEC (Dean Shipley, "FASPEC British Car Parts," *C&P*, January 1995, 67–69); German specialist Campbell-Nelson Auto Wrecking (advertisements and classifieds, *C&D*, *HVW*, and *EC*, 1985–2005); French specialist The French Car Shop (classifieds, *C&D*, 1981); Swedish specialists Swedish Auto Salvage Yard (classified, *R&T*, November 1982, 184) and Swedish Engineering (classifieds, *C&D*, 1990); and European specialists more broadly, including the yards mentioned in the text as well as N&W Foreign Auto Wrecking (advertisements and classifieds, *C&P* and *R&T*, 1971–1983).

18. Robert J. Gottlieb, "Graveyard for Classics," *MT*, April 1962, 56–57 (quotes on 57).

19. On World War II and the Korean War, see above, chaps. 1–2.

20. George A. Parks, "Perpetual Youngsters," *CL*, December 1954, 42–45; Ken Purdy, "The Vintage Car Store," *SCI*, February 1958, 26–27, 64–65; Warren Weith, untitled column, *C&D*, August 1969, 18, 20.

21. Frank Cetin, "He Brings 'Em Back to Life," *Cars*, June 1953, 52–53, 74; Russell Gerrits, "Rod's Beginning," *R&C*, June 1954, 58; Roddin' at Random, *HRM*, September 1954, 48–49; "Torrid '29 Tudor," *HRM*, December 1954, 26–27; Fred Horsley, "Building a Glass Rod," *Speed Mechanics*, April–May 1955, 8–10, 44–47; Al Franceschetti, Post Entry, *HRM*, January 1957, 10, 12.

22. Dean Shipley, "Schulte's Auto Wrecking," *C&P*, March 1996, 20–23. Because his article focuses on Schulte's son, John, Shipley does not provide the elder Schulte's first name.

23. Gottlieb, "Graveyard for Classics," 58.

24. David L. Stratton, "Museum in the Rough," *Motorsport*, August 1953, 31, 48–49 (quote on 49).

25. David L. Lewis, "The Pollard Collection," *C&P*, September 1974, 114, 114A–E.

26. "Parting out" refers to the process whereby a complete wreck is dismantled and sold piecemeal. Typically this involves dismantling a parts car owned by an individual,

although periodicals and other sources often describe salvage-yard disassembly in the same way.

27. Mark J. Hash, "Willie's Salvage and 'Museum,'" *C&P*, November 1991, 36–38, 40, 42 (quote on 36).

28. Randoll Reagan, "Charlie's Toys," *C&P*, October 2000, 52–53, 55 (quotes on 52).

29. Eric Kaminsky and Dean Shipley, "Jackson's Hudsons," *C&P*, March 1998, 22–25.

30. Ken Kittleson, "Bridley's Auto Salvage," *C&P*, July 1996, 20–22 (quote on 20).

31. Jerry W. Phillips, "Mountain Retreat," *C&P*, January 1992, 39, 44–46.

32. Joe Sharretts, "Stewart Criswell and Sons Auto Salvage," *C&P*, March 2001, 60–63 (quotes on 60 and 62).

33. Joe Sharretts, "Elwood's Auto Exchange," *C&P*, December 2006, 40–46.

34. A few examples: Robert Jay Stevens, "Easy Jack Antique Auto Parts," *C&P*, February 1982, 50–54; Joe Heacock Jr., "Hap Gemmill's Junkyard," *C&P*, May 1990, 12–14; John Robertson, "Endangered Junkyard," *C&P*, September 1990, 51–54; Dean Shipley, "Ralph's Auto Service," *C&P*, March 1993, 42–44; Dean Shipley, "Mercs and More," *C&P*, September 1995, 8–11; Eric Brockman, "Back in the Woods," *C&P*, February 1996, 38–41; Michael Ponzani, "Eperthener's Auto Wrecking," *C&P*, March 1997, 38–39; Dennis David, "Winakor's of Connecticut," *C&P*, May 1997, 42–45; Joe Sharretts, "Junkyarding in the Mountains of Pennsylvania," *C&P*, June 2005, 48–50, 52–53; Joe Sharretts, "Hoffert's Used Auto Parts," *C&P*, November 2006, 64–66, 68–69.

35. Brad Bowling, At the Wheel, *C&P*, March 2008, 6.

36. Terry Breed, Reader Forum, *C&P*, November 1998, 8.

37. Walter Lula and Mary Jane Lula, Reader Forum, *C&P*, September 1990, 162.

38. Kenneth V. Bostick, Reader Forum, *C&P*, November 1995, 167.

39. See, for example, Tony Spitoleri, Reader Forum, *C&P*, January 1982, 158; and Robert E. Smith, Reader Forum, *C&P*, August 1997, 9.

40. Keith J. Hansen, Reader Forum, *C&P*, December 1990, 155.

41. Welsh's Jaguar Enterprises advertisements and classifieds, *C&P*, *C&D*, and *R&T*, 1969–2000.

42. See for example XKs Unlimited classifieds, *R&T*, 1977–1989; British Auto Wreckers/Center classifieds, *R&T* and *C&D*, 1980–1991; Jag Parts of Arizona classifieds, *C&D*, July–December 1980; English Car Spares, Ltd., classifieds, *C&D*, 1981; Bob's Volksworld advertisements, *HVW*, 1984; Recycle Auto Parts advertisements, *HVW*, 1988; Bugs, Buses, and Things advertisements, *HVW* and *VWT*, 1994–1997; Sunray Bugs advertisements, *VWT*, 1993–1994; and Brooks Auto Service, Inc., classified, *Hemmings*, December 2003, 476.

43. Shipley, "FASPEC British Car Parts"; All Auto Parts Company classified, *R&T*, May 1977, 161; D&D Foreign Auto Dismantlers classified, *R&T*, May 1986, 182; Doc and Cy's Restoration Parts advertisements, *EC*, 1997–1998; Things Unlimited advertisements, *HVW* and *VWT*, 1997–1998.

44. Joe Sharretts, "Smith Brothers Used Auto Parts," *C&P*, October 2007, 71.

45. Leroy Drittler, "Vintage Auto Salvage," *C&P*, October 2007, 64–67; Dean Shipley, "A Lotta Auto Parts," *C&P*, December 1997, 40–43; Phil Skinner, "Walker's Garage," *C&P*, January 1997, 46–49.

46. John Robertson, "Florida Gold," *C&P*, December 1992, 40–41, 44–47.

47. Dean Shipley, "All American Classics, Inc.," *C&P*, December 1996, 43–46 (quotes on 43).

48. Tom Shaw, "Memory Lane Used Auto Parts," *C&P*, October 2005, 54–60.

49. Joe Sharretts, "Rusty's Old Iron," *C&P*, January 2008, 60–67 (quote on 60).

50. Michael G. Beda, "Speedway Auto Wrecking," *C&P*, March 2007, 54–59; Joe Sharretts, "Morrisville Used Auto," *C&P*, February 2008, 64–69; Sharretts, "Junkyarding in the Mountains," 48 (quote); Sharretts, "Albin Avenue Auto Salvage," *C&P*, July 2008, 42–49; Sharretts, "Moriches Used Auto Parts," *C&P*, May 2007, 42–47; Ken New, "Where Rust Stands Still," *C&P*, July 1990, 40–43; Randoll Reagan, "Martin Supply, Inc.," *C&P*, September 2006, 54–59; John G. Tennyson, "Gilly's Auto Wreckers," *C&P*, August 1997, 42–44. See also Dennis David, "Stewart's Used Auto Parts," *C&P*, September 1997, 42–43; and Dean Shipley, "Ernest's Auto Wrecking in Southern Colorado," *C&P*, October 1997, 42–44.

51. Bob Walton, "Unique Auto and Truck," *C&P*, July 2006, 52–57 (quote on 56); Dean Shipley, "Colorado Cache," *C&P*, December 1995, 8–11 (quote on 8); Dennis David, "Mt. Tobe Auto Parts," *C&P*, May 1999, 57–59 (quotes on 57).

52. Kevin Ross, "Vogt's Parts Barn," *C&P*, March 1990, 6–8, 10; Bob Stevens, "Papke Enterprises Might Just Be the World's Smallest Salvage Yard," *C&P*, April 1991, 58–60, 62; Stevens, "Coast G.M. Salvage," *C&P*, September 1991, 62–65; Mark J. Hash, "Oregon's Real McCoy," *C&P*, July 1992, 16–18; Dean Shipley, "East West Auto Parts," *C&P*, November 1995, 8–10, 12; Dan Abood, "Gold's Auto Wrecking," *C&P*, November 1999, 60–61, 63; Earl Duty, "Wyoming Classic Cars," *C&P*, October 2004, 48–50, 52–53; Leroy Drittler, "Vintage Auto Salvage," *C&P*, October 2007, 64–67.

53. Shipley, "A Lotta Auto Parts"; David, "Mt. Tobe Auto Parts."

54. Leroy Drittler, "R&S Auto Parts," *C&P*, August 2008, 52–59.

55. David, "Mt. Tobe Auto Parts"; Joe Sharretts, "Fredericksburg Auto Salvage," *C&P*, March 2006, 46–51; Dean Shipley, "Southwest Auto Wrecking," *C&P*, July 1995, 8–10; David, "Stewart's Used Auto Parts."

56. Joe Sharretts, "Ed Lucke Auto Parts," *C&P*, December 2007, 62–70; David, "Winakor's of Connecticut"; Bob Stevens, "A Real Graveyard for Vintage Cars," *C&P*, August 1998, 46–49, 138.

57. David, "Winakor's of Connecticut," 43.

58. Joseph J. Caro, "Treasure Trove," *C&P*, December 1990, 12–14; "Ten Best Trivial Pursuits," *C&D*, January 1988, 86–87; David, "Winakor's of Connecticut"; Tennyson, "Gilly's Auto Wreckers"; Abood, "Gold's Auto Wrecking."

59. Shipley, "Schulte's Auto Wrecking"; Ross, "Vogt's Parts Barn"; Caro, "Treasure Trove"; Richard's Auto Sales and Salvage classifieds, *C&P*, 1990–1991; Paul A. Herd, "R&R Salvage," *C&P*, June 1993, 42–43, 46.

60. Sharretts, "Criswell and Sons," 60, 62.

61. Sharretts, "Fredericksburg Auto Salvage," 48.

62. New, "Where Rust Stands Still," 40.

63. "Dog Gone Rust," *SC*, January 1997, 72. See also Dean Shipley, "Wiseman's Revisited," *C&P*, May 1996, 44–46; Desert Valley Auto Parts classified, *Hemmings*, June 2009, 454; Automania classified, *Hemmings*, November 2007, 306.

64. Bob Stevens, "They Shoot Horses, Don't They? Old Mustangs Are Corralled in this Midwestern Salvage Yard," *C&P*, August 1989, 58–60, 62; Stevens, "Papke Enterprises"; Stevens, "Visiting the 'Great Wall,'" *C&P*, May 1991, 40–42, 44, 46, 73; Stevens, "Coast G.M."

65. Dennis David, "Adler's Antique Autos, Inc.," *C&P*, June 1998, 58–59; Shipley, "A Lotta Auto Parts."

66. Rich Kimball, "Kickin' Back in Kanai," *HVW*, March 1997, 61–63.

67. Dean Shipley, "Memphis Auto Storage," *C&P*, June 1995, 8–10; Arthur Ash III, "Discover Well-Preserved Treasures at Johnson's Auto Salvage, Griffin, Georgia," *C&P*, January 2005, 50–55.

68. Doc Howell, "Idaho Surprise," *C&P*, October 1993, 18; Beda, "Speedway Auto Wrecking," 54; Shipley, "All American Classics," 43. See also Black Hills Antique Auto Barn classifieds, *C&P*, June 1977, 68, 88, 116; Hash, "Willie's Salvage and 'Museum'"; Doc Howell, "D&R Auto Sales," *C&P*, February 1992, 44–46; Bob Stevens, "Salvaging Cars in the Dakotas," *C&P*, March 1992, 16–18; Doc Howell, "Mostly Mopar," *C&P*, August 1994, 6–8; Shipley, "Mercs and More"; Shipley, "Colorado Cache"; Skinner, "Walker's Garage"; Reagan, "Martin Supply"; and Randoll Reagan, "Adams Truck and Auto Salvage," *C&P*, June 2007, 40–46.

69. New, "Where Rust Stands Still," 40.

70. See, for example, "Buy a California Chevy," *SC*, July 1981, 36–37, 68; Jeff Smith, "Treasure of Sierra Vista," *HRM*, June 1989, 135–37; Jay Jones, "356 Competition," *VW and Porsche*, November–December 1990, 30–31, 40, 100–101; Randy Moser, "K&L Auto Wrecking," *C&P*, April 1993, 64–65, 73; Randy Moser, "Larry's Auto Wrecking," *C&P*, August 1993, 8–11; Moser, "Desert Valley Auto Parts," *C&P*, June 1994, 6–9; Bob Stevens, "Memory Lane Collector Car Dismantlers," *C&P*, January 1996, 54–57; Shipley, "Wiseman's Revisited"; Stevens, "Davis Auto Wrecking of Nevada," *C&P*, April 1997, 28–30; Tennyson, "Gilly's Auto Wreckers"; and Randoll Reagan, "Hidden Treasures of the Southwest," *C&P*, July 2004, 44–46, 48.

71. New, "Where Rust Stands Still"; Moser, "K&L Auto Wrecking"; Shipley, "Wiseman's Revisited."

72. These four were the only eastern, Rust Belt, or southeastern states overrepresented in the old-car business, but Arkansas, Iowa, Kentucky, Maryland, Michigan, Minnesota, Mississippi, Missouri, New Hampshire, New Jersey, North Carolina, Pennsylvania, Ohio, Tennessee, Vermont, Virginia, West Virginia, and Wisconsin also all had slightly larger shares of the old-car business specifically than they did of enthusiast-specialty yards as a whole. By contrast, Alabama, Connecticut, Florida, Georgia, Illinois, Indiana, Louisiana, Massachusetts, New York, Rhode Island, and South Carolina all had slightly smaller shares of the old-car business vis-à-vis their shares of enthusiast yards in general. I was unable to locate a single enthusiast-oriented salvage yard in Hawaii.

73. Very few of the specialty yards I came across featured their home states in their business names, with the exception of several in Arizona. See, for example, Jag Parts of Arizona classifieds, *C&D*, July–December 1980; Arizona Desert Specialties classifieds, *C&P*, 1990–2001; Arizona Vintage Parts classifieds, *C&P*, 1992–2006; Arizona Canyon Classics classifieds, *C&P*, 1996–1998; Arizona Mailorder Autoparts classifieds,

C&D and *R&T,* August–October 1998; Z Car Source of Arizona advertisement, *Hemmings,* November 2007, 459; Arizona Auto Wrecking classifieds, *C&P,* 1999–2000; Arizona Vintage Parts advertisements, *Hemmings,* 2007–2009; and Arizona Muscle Mart advertisement, *Hemmings,* 2007–2008. For some examples of Rust Belt, northeastern, and East Coast yards that sold Arizona parts, see Stevens, "Bill Lindsay's Ford Collection"; Murray Park advertisement, *Hemmings,* August 2008, 234; Desert Dog Auto Parts advertisement, *Hemmings,* November 2007, 327; Boos-Herrel advertisement, *Hemmings,* November 2007, 573; and Jim's Auto Center advertisement, *OCW,* 7 May 2009, 28.

74. Bob Stevens, "Recycling Porsches & Bimmers," *C&P,* July 1991, 56–58; Stevens, "Coast G.M."; Phil Skinner, "Renegade Wrecker, Law Abiding Citizen," *C&P,* December 1999, 56–57, 59.

75. One does wonder what the old-car salvage business might have looked like if California's share were skewed statistically to the same extent as Arizona's.

76. Heacock, "Hap Gemmill's Junkyard"; In the Headlights, *C&P,* April 1991, 189; Bob Stevens and Dean Shipley, Salvage Yard Alert, *C&P,* November 1993, 77; Dean Shipley, "Swartz Salvage Sale," *C&P,* July 1994, 67, 71–72; Bob Stevens, Salvage Yard Alert, *C&P,* January 1998, 142; Peterson, "Leo Winakor & Sons"; Phil Skinner, "Harmon Auto Wrecking," *C&P,* January 2001, 44–45, 47; In the Headlights, *C&P,* April 2005, 10; Joe Sharretts, "Wheels of Time Salvage Yard, Part 1," *C&P,* September 2008, 64–71; Ron Kowalke, "Hauf Closing 63-Year-Old Oklahoma Salvage Business," *OCW,* 2 July 2009, 22.

77. Paraphrased in Kowalke, "Hauf Closing."

78. On Clark, see Stevens and Shipley, Salvage Yard Alert. On local officials and old-car-yard closures, see Joseph J. Caro, "Welcome to Glen's," *C&P,* May 1990, 28–30; Skinner, "Harmon Auto Wrecking"; In the Headlights, *C&P,* April 1991, 189; and, for an overview, Bob Stevens, At the Wheel, *C&P,* August 1989, 69.

79. Stevens, At the Wheel; Bob Swett and Elsie Swett, Reader Forum, *C&P,* August 2004, 9–10; Todd Seaward, "Saying Goodbye," *C&P,* January 2006, 62–66; Sharretts, "Rusty's Old Iron"; Ron Kowalke, "Mississippi Crush Is Not a Dance Nor a Drink," *OCW,* 25 June 2009, 9, 12 (quote on 9); Kowalke, "Mississippi's Last Holdout," *OCW,* 9 July 2009, 38–39.

80. Walter Firey, "Ecological Considerations in Planning for Rurban Fringes," *American Sociological Review* 11 (August 1946): esp. 416; Robert H. Nelson, *Zoning and Property Rights: An Analysis of the American System of Land-Use Regulation* (Cambridge, Mass., 1977), 1–2; William A. Fischel, *The Economics of Zoning Laws: A Property Rights Approach to American Land Use Control* (Baltimore, 1985), 25–27; John O'Looney, *Economic Development and Environmental Control: Balancing Business and Community in an Age of NIMBYs and LULUs* (Westport, Conn., 1995), 1–4; Daniel W. Bromley, "Property Rights and Land Use Conflicts: Reconciling Myth and Reality," in *Economics and Contemporary Land Use Policy: Development and Conservation at the Rural-Urban Fringe,* ed. Robert J. Johnston and Stephen K. Swallow (Washington, D.C., 2006), 38–51; Heacock, "Hap Gemmill's Junkyard"; Caro, "Welcome to Glen's"; Robertson, "Endangered Junkyard"; Dean Shipley, "Salvage Yard Alert," *C&P,* January 1994, 81; Dean Shipley, "Salvage Yard to Be Auctioned," *C&P,* February 1994, 18;

Bayman Auction Service advertisement, *C&P*, April 1994, 83; Sharretts, "Wheels of Time."

81. Sharretts, "Smith Brothers," 68. See also Parks C. Carpenter, Reader Forum, *C&P*, April 1984, 62; Richards Sisson, Reader Forum, *C&P*, July 1990, 170; Peterson, "Leo Winakor & Sons"; Bob Swett and Elsie Swett, Reader Forum.

82. Spence Murray, "Hot Rod Body Shells of Glass," *PHR*, May 1963, 70–76; Steve's Auto Restorations, Inc., advertisement, *Hemmings*, December 2003, 420; "Bench Racing," *HRM*, December 2004, 18; Real-Deal Steel advertisement, *PHR*, May 2011, 67.

4. Junkyard Jamboree

1. On early periodicals and salvage-yard coverage, see above, chaps. 1–2.

2. "Repairs versus Replacements," *Motor Thrift*, 1956 (annual), 70–71; Joe Petrovec, "Shop JUNKYARDS and Save," *Auto Age*, February 1957, 20–23; Alex Walordy, "Replacement Parts for Your Car," *CL*, September 1958, 32–33, 60–63.

3. Don Francisco, "How to Buy Used Engines," *CC*, February 1957, 12–17, 58–59; Jack Phelps, "Bargains in Horsepower," *ML*, November 1955, 26–27, 66; Orin Tramz, "10 Steps to Customizing," *R&C*, October 1953, 28–33; Felix Zelenka, "Torch Tips," *CC*, December 1953, 24–27; Russ Kelly and Len Griffing, "Wrecking Yard Bit," *SCI*, February 1958, 30–31.

4. See above, chap. 2.

5. On the practice of hot rodding and customizing in the 1920s–1950s, see above, chap. 2.

6. Willard C. Poole, "Old Car Treasure Hunt Is On!" *Cars*, May 1953, 36–37, 79 (quote on 36).

7. Hank Wieand Bowman, "Sutton's Shiny Scraps," *Auto Age*, October 1954, 38–39, 46 (quote on 46).

8. A few did hint at the experience of "finding treasure" in the 1970s, but rarely in precisely those words. See, for example, Menno Duerksen, "We've Been to Hershey '70," *C&P*, November 1970, 58–67; Miscellany, *C&P*, August 1971, 98; and David L. Lewis, "Pollard Collection," *C&P*, September 1974, 114, 114A–E. For some examples from the 1980s, see Rich Kimball, Collector's Corner, *HVW*, October 1982, 84, 86, 90, 92, 94; Jeff Walters, "Treasure Hunting in Jamaica," *HVW*, December 1985, 96–99, 138, 140; James G. Watts, Reader Forum, *C&P*, June 1982, 161; Peter Egan, Side Glances, *R&T*, March 1989, 20, 24; and Jeff Smith, "Treasure of Sierra Vista," *HRM*, June 1989, 135–37.

9. Joseph Diorio, The Letter Box, *Motorsport*, June–July 1955, 5.

10. See, for example, Bowman, "Sutton's Shiny Scraps."

11. See above, chap. 2, and David N. Lucsko, *The Business of Speed: The Hot Rod Industry in America, 1915–1990* (Baltimore, 2008), 228.

12. Tom Medley, "Chicken Coop Tudor," *R&C*, 1977, 20–21; "Larry Wood, 1932 Nash Brougham," *R&C*, April 1973, 17–18, 20–21; "Reincarnation," *PHR*, May 1976, 54–56; "Hot Rod Gallery," *HRM*, October 1978, 54.

13. Vintage Tin, *HRM*, July 1971, 122.

14. "Junkyard Jamboree," *HRM*, February 1977, 70–74, 76–83; Gray Baskerville, "Rod, Radicalize, or Restore," *R&C*, April 1973, 67–71.

15. See, for example, Vintage Tin, *R&C*, September 1972, 33; Vintage Tin, *R&C*, January 1974, 64; and Early Iron, *Street Rodder*, June 1989, 146. See also Vintage Tin, *HRM*, July 1971, 122, where the column's title appears alongside a cartoon of a man with a metal detector wandering in a field near an abandoned antique car.

16. "Solution to Roadster Body Shortage," *HRM*, September 1952, 26–29; Fletcher Hines, "Plastic Bodies for Home Production," *Hop Up*, April 1953, 14–15, 17.

17. Tom Bates, "The Glass Menagerie," *HRM*, October 1962, 76–81.

18. Lucsko, *Business of Speed*, chap. 9.

19. Rebuildable engines from the 1960s did remain widely available; see, for example, Jim Losee's "Salvage Yard Treasure Hunt" series, which ran in *Popular Hot Rodding* during 1984 (e.g., February, 46–48; May, 36–39; and June, 62–65).

20. Lucsko, *Business of Speed*, 223–25.

21. See John Thawley, "Wreck-Connaissance," *R&C*, May 1974, 24–26, 37–39; Cam Benty, "The Salvage Yard Maze," *PHR*, August 1982, 76–79; and Losee's "Salvage Yard Treasure Hunt" series.

22. Duerksen, "We've Been to Hershey '70," 59.

23. Carl Zimring, *Cash for Your Trash: Scrap Recycling in America* (New Brunswick, N.J., 2005), 21, 126–27.

24. *Cars and Parts* in particular often used the less offensive "salvage yard," but, like *Hot Rod, Rod and Custom, Hot VWs*, and the rest, it often used "junkyard" too.

25. Menno Duerksen, "Hershey 74," *C&P*, November 1974, 124–28 (quotes on 124 and 125).

26. John Thawley and Gray Baskerville, "Swap Meet Price Comparisons," *R&C*, September 1972, 18.

27. Kevin Clemens, On the Line, *EC*, September 2002, 8; Matthew King, "One Man's Trash," *CC*, July 2000, 98. See also Peter Egan, Side Glances, *R&T*, March 1989, 20, 24; and Brock Yates, "You May Already Be Rich," *C&D*, July 1997, 20. Section title quote from Pittsburgh Parts-A-Rama advertisement, *SS*, May 1983, 75.

28. See above, chap. 3.

29. Bob Stevens, "Pinky's Cars," *C&P*, April 1997, 44–47 (quote on 46).

30. Ken New, "Vernon Burks' Model A Fords," *C&P*, July 2004, 72–75.

31. Lee Beck, "Carolina Collectors," *C&P*, November 1990, 38–41 (quote on 38).

32. Henry Platt, "Jackpot," *HRM*, May 2009, 38–44, 46 (quote on 38).

33. Cam Benty, "American Addiction," *HRM*, July 2008, 30–32, 34–36, 38–42, 44; Spike, quoted in Hidden Treasure of the Month, *HRM*, July 2004, 18. See also Roger Johnson, "Bow-Tie Treasures," *HRM*, July 1992, 80–81, 83.

34. John Pearley Huffman, "Musclecar Jackpot," *HRM*, August 2006, 48–54, 56 (quotes on 49).

35. Dean Shipley, "Parrie's Prizes," *C&P*, May 1992, 62–64 (quote on 64).

36. Jim Benjaminson, "Robert Glass Collection Dispersal," *C&P*, September 1996, 34.

37. See, for example, Melvin Powers Auctioneer advertisement, *C&P*, May 1982, 165; In the Headlights, *C&P*, April 1990, 74; Roger L. Neal Auctioneer advertisement, *C&P*, October 1993, 158; and Barry and Slosberg, Inc., Auctioneers, advertisement, *C&P*, February 2008, 12.

38. "Cache Catch," *HRM*, March 1982, 18–19. Automobile Atlanta, a Porsche specialist, made a similar purchase of NOS parts in the early 1990s (advertisement, *EC*, January 1993, 6).

39. See, for example, Frey & Sons Auctioneers advertisement, *C&P*, August 1989, 84; Merle Clark Auctioneer advertisement, *C&P*, May 1992, 148; and Kruse International advertisement, *C&P*, March 2000, 70.

40. Buick Farm classified, *C&P*, December 1990, 88; Herbie's VW Hideaway advertisements, *HVW* and *VWT*, 1996–1998; Sandy Esslinger, "Shop Tour," *VWT*, February 1997, 62–63; Baxter Ford Parts advertisement, *Hemmings*, December 2003, 422.

41. Rich Kimball, Collector's Corner, *HVW*, September 1982, 90, 94, 96, 98; Kimball, Collector's Corner, *HVW*, October 1982, 84, 86, 90, 92, 94; Jeff Walters, "Treasure Hunting in Germany," *HVW*, October 1984, 62–64, 103; Walters, "Treasure Hunting in Jamaica"; Walters, "Treasure Hunting in Guatemala, Part 1," *HVW*, September 1987, 88–91, 128, 132; Walters, "Treasure Hunting in Guatemala, Part 2," *HVW*, October 1987, 46–49, 139; Walters, "Treasure Hunting in Venezuela," *HVW*, September 1988, 44–47. See also Rich Kimball, "Treasure, What Treasure? The Saga Continues," *HVW*, May 1996, 50–51; and Kimball, Collector's Corner, *HVW*, June 2004, 30.

42. Walters, "Treasure Hunting in Germany," 62.

43. Walters, "Treasure Hunting in Guatemala, Part 2," 49.

44. Kimball, Collector's Corner, September 1982, 94.

45. Kimball, Collector's Corner, September 1982 and October 1982.

46. Bob Stevens, "Sweet Treats," *C&P*, January 1994, 52–56, 58; "Hershey Fall Meet," *Automobile*, September 2004, special insert, 5; Little Hershey Swap Meet and Car Show advertisement, *C&P*, April 1982, 166; 7th Annual Kyana Region Antique Automobile Club of America advertisement, *C&P*, January 1975, 97.

47. Mansfield Trading Bee advertisement, *C&P*, April 1992, 160; Eighth Annual Mid-Winter Kalamazoo, Michigan, Sell and Swap advertisement, *C&P*, November 1970, 48; Eric Kaminsky, "Portland Swap Meet," *C&P*, August 1997, 56–57; Robert Jay Stevens, "California Swappin' at the Rose Bowl," *C&P*, January 1982, 62–63; Hoosier Auto Show and Swap Meet advertisement, *C&P*, July 1969, 35; Miscellany, *C&P*, October 1975, 3; Southwest Swap Meet advertisement, *C&P*, August 1989, 163; Walt Reed, "26th Annual Fall Barrie," *C&P*, January 1997, 134; Reed, "Barrie Automotive Flea Market," *C&P*, September 1997, 40; Carlisle Swap Meet Thank-You advertisement, *C&P*, December 1974, 45.

48. Southwest Swap Meet advertisement; Kaminsky, "Portland Swap Meet"; Hoosier Auto Show and Swap Meet advertisement.

49. See, for example, Twelfth Annual Abilene, Texas, Swap Meet and Flea Market advertisement, *C&P*, January 1975, 94; Bob Stevens, "Turlock '96," *C&P*, May 1996, 37–38; Roadster Exhibition, Trade Show, and Swap Meet advertisement, *SS*, March 1980, 17; 14th Annual Carnival of Cars Flea Market and Car Show advertisement, *C&P*, June 1977, 139; and Car Custom Bug-O-Rama XXVII advertisement, *HVW*, June 1991, 104.

50. John Robertson, "Spring Carlisle '92," *C&P*, July 1992, 10, 12–13 (spring Carlisle events were added in 1977); Greg Rager, "Spring Carlisle 2007," *C&P*, August 2007, 64–66.

51. Antique Auto Show advertisement, *C&P*, April 1974, 113; Giant Kyana Region AACA 6th Annual Swap Meet advertisement, *C&P*, April 1974, 119 ("junque"); Annual Southern California Antique and Classique Auto Auction and Trade Fair advertisement, *C&P*, July 1975, 33 ("classique").

52. Bob Lichty, "Pricing Hershey," *R&C*, March 1974, 58–59.

53. Duerksen, "Hershey 74," 124.

54. Thawley and Baskerville, "Swap Meet Price Comparisons"; Lichty, "Pricing Hershey"; Rich Kimball, Collector's Corner, *HVW*, December 1983, 22 (quote), 94, 96.

55. Steve Magnante, "Bottom Feeders," *HRM*, February 2003, 88–91; Robert J. Gottlieb, Classic Comments, *MT*, October 1964, 96–97; Michael Lamm, "A Day at Harrah's," *MT*, November 1965, 130–32; Smith, "Dream-Come-True Oval"; Arch Brown, "1957 Oldsmobile Starfire 98," *C&P*, June 2000, 20–27; "Basement Gold," *CHP*, February 1993, 59–62; Ken New, "Pate Swap Meet Celebrates 20th," *C&P*, August 1992, 72–73.

56. Bob Stevens, "Barn-Fresh Finds," *C&P*, October 2004, 40–41 (quote on 41).

57. See above, chap. 2. The section heading echoes a common gearhead cry; see, for example, F. A. Wright, Reader Forum, *C&P*, December 1991, 153.

58. See above, chap. 2.

59. From the Four Winds, *MA*, 6 June 1912, 49; Dorothy B. Nichols, "'A Little Old Last Year's Car' That's Almost Forgotten," *MA*, 6 January 1916, 40–41.

60. "Dealer Takes in a Real Old Timer," *FD&SF*, October 1930, 38.

61. Poole, "Old Car Treasure Hunt," 37.

62. Ken Purdy, "Vintage Car Store," *SCI*, February 1958, 26–27 (quotes).

63. Miscellany, *C&P*, August 1971, 98.

64. Thos L. Bryant, "1915 FRP Touring Car Discovered," *R&T*, June 1976, 106, 108.

65. Members' Machines, *SS*, May 1985, 50.

66. Don Kear, Reader Forum, *C&P*, January 1984, 152. Kear reported that he found a 1935 Cord, which if true would mean it was a rare preproduction car from 1935. Much more likely, what he saw was a 1936 or 1937 model.

67. "Antique Cars Found," *AI*, 1 September 1971, 121.

68. Herbert W. Hesselmann and Halwart Schrader, *Sleeping Beauties* (Zurich, 2007), 5.

69. Randy Slotten, Reader Forum, *C&P*, March 1984, 60; Robert K. Smith, "Above-Standard Standard," *HVW*, August 1966, 40–43; "Middle-Age Madness," *HRM*, July 1996, 44–45; Eric Meyer, "It's a *What*? Rometsch's Beeskow Is No Porsche, but It *Is* One of Several Coachbuilt VWs that Competed Directly with Ferry's Idea of a Post-War Sports Car," *Excellence*, November 2004, 70–76; Bob Mehlhoff, "Somewhere in Time," *CHP*, May 2001, 106, 108, 110; Phil Hill, "1953 Ferrari 375 MM," *R&T*, October 1997, 144–47, 150; "Barn Finds," *HRM*, October 2007, 48; "Packard Rolls Out into the Open," *C&SC*, January 2011, 26; "Barn-Again DB Aston," *C&SC*, March 2011, 18.

70. The site carsinbarns.com debuted in 1999, for example (www.carsinbarns.com/siteinfo.html, accessed 10 February 2015). *Chasing Classic Cars* debuted in 2008, and *What's in the Barn?* in 2013.

71. The others: *The Vincent in the Barn* (2009), *The Corvette in the Barn* (2010), and *The Harley in the Barn* (2012), all published by Motorbooks International in Minneapolis.

See also Michael E. Ware, *Automobiles Lost and Found: Extraordinary Stories of Long-Lost Cars Rediscovered* (Newbury Park, Calif., 2008).

72. Michael T. Lynch, "1955 Ferrari 121 Le Mans Scaglietti Roadster," *R&T*, March 1998, 98–101, 104 (quote on 104).

73. David Featherston, "Foundling," *EC*, June 1993, 110–12 (quotes on 111).

74. Robert K. Smith, "Renewed '52," *HVW*, April 1997, 76–79 (quotes on 78).

75. "Custom Cars," *HRM*, June 1986, 84–86 (quote on 84). Sam Barris's brother, George, was also a legendary customizer in the 1950s and 1960s.

76. Pat Ganahl, "The Orbitron Saga," *HRM*, March 2008, 70–76, 78 (quote on 78).

77. Jerry Heasley, "The Mexican Connection," *PHR*, October 1991, 58–60; "How to Build a Pro Street/Pro Gas '67 Nova SS," *SC*, October 1981, 48–49; "Hardwood Hot Rod," *HRM*, January 1991, 74–75; Damon Lee and Dave Hill, "Super Chevy Three Pack," *SC*, December 1999, 76–78; Carlton Carpenter, Hidden Treasure of the Month, *HRM*, March 2005, 18.

78. Examples of junkyard-scene artists: Trenton Frost advertisement, *C&P*, May 1975, 107; Vintage Varieties advertisement, *C&P*, July 1975, 96; Larry Crane, "Non Sequitur," *Automobile*, September 1996, 118–23; Dale Klee Studio advertisement, *Hemmings*, December 2003, 41. Video tours of salvage yards: Old Car City advertisement, *C&P*, August 1989, 82; Junkyard Junkies advertisement, *C&P*, November 1991, 141; Steve Drake classified, *C&P*, December 1992, 76; J. Miller Restorations classified, *C&P*, December 1994, 107; Video Salvage advertisement, *C&P*, December 1995, 150. Examples of websites featuring derelict and junkyard cars: www.junkyarddog.com/oldcars.htm and www.oldride.com/rustyrides.html (both accessed 21 July 2011). Photograph-rich books on automotive junk: Pat Kytola and Larry Kytola, *Diamonds in the Rough: American Junkyard Jewels* (Minneapolis, 1989); *Cars and Parts* editorial staff, *Salvage Yard Treasures of America* (Sydney, Ohio, 1999); Will Shiers, *Roadside Relics: America's Abandoned Automobiles* (Minneapolis, 2006). *Hemmings* has long produced salvage-yard calendars.

79. Some examples: www.flickr.com/photos/detroitderek/sets/72157601861458499/ (accessed 21 July 2011); www.detroityes.com/home.htm (accessed 21 July 2011); detroiturbex.com (accessed 13 November 2013); www.forbidden-places.net/urban-exploration-gary-indiana-ghost-town (accessed 13 November 2013). See also the "history" section at www.englishrussia.com (accessed 9 December 2013).

80. William Jeanes, "Ten Best Awful Automotive Relics," *C&D*, January 1993, 59–61.

81. Fertig, *VWT*, October 1993, 118; May 1994, 106; December 1994, 107; September 1995, 109; and February 1998, 86.

82. Hidden Treasure of the Month, *HRM*, July 2005, 22; Hidden Treasure of the Month, *HRM*, December 2006, 23; Hidden Treasures of the Month, *HRM*, January 2005, 20; Hidden Treasure of the Month, *HRM*, August 2006, 25.

83. Bob Stevens, "Rare Edsel Wagon Is Roadside Landmark," *C&P*, May 2006, 10; "Weathered Wheels," *OCW*, 4 June 2009, 20; Money Shot, *EC*, May 2010, 14–15; "Ferrari Fixer-Upper," *C&D*, January 1991, 34; "Best of 2010," *C&SC*, February 2011, 12; "Goodbye Lord, I'm Going to Bodie," *PHR*, June 1995, 11.

84. Bill McGuire, "The Ghosts of Detroit," *HRM*, October 2009, 70–77.

85. Dan Burkholder, "The Left-Behinds," *HVW*, September 1991, 58.

86. PS, *R&T*, May 1990, 178; George Harnden, Letter of the Month, *HRM*, July 1999, 8; "Department of Sad Fates Update," *CC*, August 2000, 10; Steve Magnante, "Mali Boo-Boo," *HRM*, April 2002, 19.

87. "Vintage Racing Takes a Big Hit," *C&D*, October 2005, 38; Wrecked Exotics, *MPH*, October 2005, 17; In the Headlights, *C&P*, November 2005, 13.

88. Bob Tyler, Reader Forum, *C&P*, May 1998, 9; Wreck of the Week, *OCW*, 25 June 2009, 11, 2 July 2009, 11, 9 July 2009, 22.

89. Reader Sightings, *C&D*, May 1997, 14, and March 1998, 14.

90. PS, *R&T*, May 1983, 224, and April 1991, 183.

91. People and Places, *R&T*, August 1982, 4. See also "Assault," *R&T*, November 2005, 34, a similar story from Germany.

92. Dale Kruti, cartoon, *HRM*, April 1967, 137; cartoon, *PHR*, October 1972, 14; George Ziegler, cartoon, *SS*, August 1987, 12; George Trosley, Trosley's Page, *VWT*, November 1991, 92; "Frank & Troise," *R&T*, May 1996, 17; "Wheelspin," *R&T*, July 2008, 24.

93. Wrecked Exotics, *MPH*, June–July 2005, 22, August 2005, 23, October 2005, 17, November 2005, 27, and February 2006, 20.

94. David L. Lewis, Ford Country, *C&P*, April 1990, 64.

95. See, for example, Bob Stevens and Dean Shipley, Salvage Yard Alert, *C&P*, November 1993, 77; Dean Shipley, Salvage Yard Alert, *C&P*, January 1994, 81; Bob Stevens, Voice of the Hobby, *C&P*, December 1995, 76; Salvage Yard Alert, *C&P*, June 1997, 58; and Stevens, Salvage Yard Alert, *C&P*, January 1998, 142.

96. See, for example, Bob Stevens, "The 'Graveyard' Is Sold," *C&P*, March 1982, 178–79; Joe Heacock Jr., "Hap Gemmill's Junkyard," *C&P*, May 1990, 12–14; John Robertson, "Endangered Junkyard," *C&P*, September 1990, 51–54; Todd Seaward, "Saying Goodbye," *C&P*, January 2006, 62–66; and Joe Sharretts, "Smith Brothers Used Auto Parts," *C&P*, October 2007, 68–72. See also Street Sweeper, *SS*, August 1984, 70.

97. See, for example, Laurence Suhsen, Reader Forum, *C&P*, January 1984, 150; Slotten, Reader Forum; Michael A. Berry, Reader Forum, *C&P*, March 1984, 176; Keith J. Hansen, Reader Forum, *C&P*, December 1990, 155; Wright, Reader Forum; and Andy Williams, Reader Forum, *C&P*, February 2007, 6.

98. Hidden Treasure of the Month, *HRM*, August 2004, 20.

99. Duff Gray, Hidden Treasures of the Month, *HRM*, January 2005, 20. See also Hidden Treasure of the Month, *HRM*, July 2006, 26, December 2006, 23, and February 2009, 15. For earlier examples, see H. L. Stutts Jr., Post Entry, *HRM*, September 1981, 4; and Richard Nesthus, Letter of the Month, *HRM*, October 1999, 12.

100. See, for example, "Finders Keepers," *R&C*, April 1968, 62; Vintage Tin, *R&C*, September 1972, 33; Vintage Tin, *R&C*, January 1974, 64; Jim Shane, "I See Dead Cars," *C&D*, June 2003, 92–94, 96, 99; Greg Carr, Bug Mail, *HVW*, December 1994, 9; Ran When Parked, *Classic Motorsports*, March 2010, 23, 25; and "Olympic Will Run Again," *C&SC*, January 2011, 24.

101. Peter Egan, Side Glances, *R&T*, November 1988, 24.

102. David Freiburger, "Patina," *HRM*, April 2007, 60–66 (quote on 61). See also Richard Bremner, "Left Well Alone, or Barn Again? Having Sunk His Savings into a

Highly Original—But Very Rotten—Austin Mini, Richard Bremner Faces the Eternal Dilemma: To Restore or to Preserve," *C&SC*, August 2011, 110–13, 115.

103. Fertig, *VWT*, November 1993, 118.

104. Steve Magnante, "Hashed Hemi," *HRM*, December 2006, 37.

105. Badger Brass Manufacturing Company advertisement, *MA*, 12 October 1905, 44; Pyrene Manufacturing Company advertisement, *Motor*, May 1913, 176; American Collectors Insurance advertisement, *C&P*, May 2008, 41; Ford Motor Company and the Governors Highway Safety Administration advertisement, *C&D*, March 2003, 119.

106. Daniel Strohl, "Light a Match, Will Ya?" *Hemmings*, November 2007, 86.

107. Robert Jay Stevens, "Easy Jack Antique Auto Parts," *C&P*, February 1982, 50; Rob Kinnan, "Hole Shot," *HRM*, February 2007, 14.

108. "Ten Best Trivial Pursuits," *C&D*, January 1988, 86–87.

109. Frederick Inocencio Collazo, "Volkswagens in the Yard," *VWT*, February 1998, 47–51 (quote on 48); Peter Egan, Side Glances, *R&T*, December 1999, 29–31 (quote on 31).

110. Kevin Clemens, On the Line, *EC*, June 2001, 6, 8.

111. Heacock, "Hap Gemmill's Junkyard: Old Car Collectors Must Play 'Beat the Crusher'"; Robertson, "Endangered Junkyard"; Ryan Lee Price, "The Savior of the Type II: Transporters Find Sanctuary with Rich Yost," *VWT*, December 2003, 80–83; Salvage Yard Alert, *C&P*, June 1997, 58.

112. On "de facto junkyards," see below, chap. 5.

5. Not in My Neighbor's Backyard, Either

1. Eric Kaminsky, Voice of the Hobby, *C&P*, November 1999, 60; West Peterson, Voice of the Hobby, *C&P*, February 2000, 43. See also "Martic Moving Forward on Hess Property Cleanup," *Intelligencer Journal/Lancaster New Era* (Pennsylvania), 13 July 2005, which includes a retrospective look at the Groff case.

2. On Lady Bird's Bill, see Carl Zimring, *Cash for Your Trash: Scrap Recycling in America* (New Brunswick, N.J., 2005), chap. 5; Zimring, "Neon, Junk, and Ruined Landscape: Competing Visions of America's Roadsides and the Highway Beautification Act of 1965," in *The World beyond the Windshield: Roads and Landscapes in the United States and Europe*, ed. Christof Mauch and Thomas Zeller (Athens, Ohio, 2008), 94–107; Zimring, "The Complex Environmental Legacy of the Automobile Shredder," *Technology and Culture* 52 (July 2011): 523–47; Tom McCarthy, *Auto Mania: Cars, Consumers, and the Environment* (New Haven, Conn., 2007), chap. 8; and above, chap. 1.

3. In *Chasing Dirt: The American Pursuit of Cleanliness* (New York, 1995), Suellen Hoy convincingly links postwar America's obsession with clean and up-to-date automobiles to a broader fixation on cleanliness and sanitary living. Implicit in this link is a disdain for things, especially cars, that are old or dirty. Lady Bird's efforts appear to have meshed with this attitude to generate the anti-ratty-looking-car fervor of the 1960s and beyond. See also Jonathan Sterne, "Out with the Trash: On the Future of New Media," in *Residual Media*, ed. Charles R. Acland (Minneapolis, 2007), esp. 27, for a discussion of the role of the state in maintaining a "mythically pure" public aesthetic.

4. Willie, Douglas Harper's subject in *Working Knowledge: Skill and Community in a Small Shop* (Berkeley, Calif., 1987), exemplified this too.

5. Philip G. Zimbardo, "A Field Experiment in Auto Shaping," in *Vandalism*, ed. Colin Ward (London, 1973), 85–90; Larry Ford and Ernst Griffin, "The Ghettoization of Paradise," *Geographical Review* 69 (April 1979): 140–58; Kevin Lynch with Michael Southworth, *Wasting Away: An Exploration of Waste* (San Francisco, 1990), 62–63; Zimring, *Cash for Your Trash*, 107; McCarthy, *Auto Mania*, 86, 153–55.

6. Licensing: Bob Stevens, "Coast G.M. Salvage," *C&P*, September 1991, 62–65; Lynne K. Varner, "Junkyard Stirs Ire of Neighbors," *WP*, 9 April 1992; Karen A. Holness, "DEEDCO Fights Plan to Build Junkyard," *Miami Times*, 19 June 1997; Dana Hedgpeth, "Knowing What's Hot Is the Key to Auto Recycling," *WP*, 28 August 2006. Nuisance and zoning: Robert H. Nelson, *Zoning and Property Rights: An Analysis of the American System of Land-Use Regulation* (Cambridge, Mass., 1977), 1–2; William A. Fischel, *The Economics of Zoning Laws: A Property Rights Approach to American Land Use Control* (Baltimore, 1985), 25–27; John O'Looney, *Economic Development and Environmental Control: Balancing Business and Community in an Age of NIMBYs and LULUs* (Westport, Conn., 1995), 1–4.

7. Examples abound. For a taste see LaBarbara Bowman, "City Searching for Way to Redevelop Run-Down Areas of Buzzard Point," *WP*, 31 March 1977; Martha M. Hamilton, "Salvage Tried on Harambee Area," *WP*, 25 September 1980; and "Scrap the Junk and Save the Street," *WP*, 27 September 1980 (projects in the D.C. area). See also Ron Brownlow, "Amid Dents, Diamonds," *NYT*, 29 May 2005; Donald Bertrand, "City Wants Plans for 'Triangle,'" *NYDN*, 13 February 2006; Terry Pristin, "Home Is Where the Auto Parts Are," *NYT*, 17 September 2006; Michael Saul, "No Way Am I Gonna Leave Junk Heaven," *NYDN*, 2 May 2007; Fernanda Santos, "A Dilapidated Tract of Queens, and a Fight to Control Its Future," *NYT*, 9 September 2008; and "Junk the Junkyards," *NYDN*, 25 September 2008 (plans to redevelop the Willets Point triangle in Queens, New York).

8. Ronald D. White, "Alexandria Junkyard Fighting for Its Life," *WP*, 13 March 1981; Curtis Krueger, "Developers Want Piece of Pinellas' Last Frontier," *SPT*, 14 March 1988; Thomas C. Jorling, Commissioner, Environmental Conservation, State of New York, "Syracuse Starts Long Road Back from Environmental Blight," *NYT*, 8 November 1990; Leslie Eaton, "Syracuse Looking to Mall for Tourists (and Money)," *NYT*, 13 August 2000; Jonathan Miller, "Lipstick on a Pig," *NYT*, 18 July 2004; Donna Walter, "77-year-old Junkyard Owner Fights Her Way to Supreme Court," *Daily Record and the Kansas City Daily News-Press*, 11 October 2007.

9. On "rurban" areas, see Walter Firey, "Ecological Considerations in Planning for Rurban Fringes," *American Sociological Review* 11 (August 1946): 411–23; see also Daniel W. Bromley, "Property Rights and Land Use Conflicts: Reconciling Myth and Reality," in *Economics and Contemporary Land Use Policy: Development and Conservation at the Rural-Urban Fringe*, ed. Robert J. Johnston and Stephen K. Swallow (Washington, D.C., 2006), 38–51.

10. Robert Bruegmann, *Sprawl: A Compact History* (Chicago, 2005), esp. chaps. 4–5; compare with Kenneth T. Jackson, *Crabgrass Frontier: The Suburbanization of the United States* (New York, 1985).

11. See, for example, Nelson, *Zoning and Property Rights*; Fischel, *Economics of Zoning Laws*; O'Looney, *Economic Development and Environmental Control*; Herbert Inhaber,

Slaying the N.I.M.B.Y. Dragon (New Brunswick, N.J., 1998); and Gregory McAvoy, *Controlling Technocracy: Citizen Rationality and the NIMBY Syndrome* (Washington, D.C., 1999). See also Bromley, "Property Rights and Land Use Conflicts," which delves into Locke and Kant to argue that property rights are not inherently permanent, and thus it does not matter which came first, the nuisance or the backyard, because neither has an absolute moral or legal right to the land. None of the actors discussed in this chapter framed their disputes in these terms. Instead, all of their arguments implied that property rights are indeed absolute; where they differed was the question of what those rights allow.

12. Fischel, *Economics of Zoning Laws*, 284–90 (farms); Daniel Simone and Kendra Myers, "Weekend Warriors: The Survival and Revival of American Dirt-Track Racing," in *Horsehide, Pigskin, Oval Tracks and Apple Pie: Essays on Sports and American Culture*, ed. James A. Vlasich (Jefferson, N.C., 2006), 136–37 (racetracks); Betsy Braden and Paul Hagan, *A Dream Takes Flight: Hartsfield Atlanta International Airport and Aviation in Atlanta* (Athens, Ga., 1989), 239n14 (airports).

13. Regional Report, *WTG*, 17 October 1989. See also David Winters, "Town of Oswegatchie Eyes Higher Junkyard Fees," *Watertown Daily Times* (New York), 23 December 2007; and Patrick Lester, "Bucks Man Faces Trouble over Illegal Junkyard," *Morning Call* (Allentown, Pa.), 29 March 2008.

14. Dan Benson, "Junkyard Protests 'Deluge' of Citations," *MJS*, 20 August 2007; "Permit Approved for Kewaskum Salvage Yard," *MJS*, 22 November 2007.

15. Jeff Reinitz, "Closed-Down Junkyard Operators Take Waterloo, Iowa, to Court over Licensing," 25 August 2004; Tim Jamison, "Waterloo Salvage Yard Stirring Neighbors Again," 12 August 2007; "Negotiations May End in Salvage Yard Move," 28 March 2010; and "Waterloo Council, Salvage Yard Reach Buyout Terms," 22 June 2010, all in *Waterloo–Cedar Falls Courier* (Iowa).

16. Collins Conner, "Junkyard Has Heaps of Trouble," *SPT*, 14 March 1991; Conner, "Junkyard Owner Mends Fence, Sort Of," *SPT*, 6 April 1991; Jeffrey S. Solochek, "It's Cleanup Time at the O.K. Corral in Hernando County," *SPT*, 7 June 2000; Solochek, "County Goes to Court over Junk," *SPT*, 11 August 2000; Solochek, "Injunction Limits Options Available to Junk Dealer," *SPT*, 12 October 2000; Solochek, "County Again Pledges to Clean Up Junkyard," *SPT*, 22 January 2002; Jennifer Liberto, "Code Officers Win Battle for Hubcap City," *SPT*, 13 July 2004. For a defense of the O.K. Corral in light of its tacky neighbors, see Jan Glidewell, "Junkyard Is Just One of the Eyesores on U.S. 19," *SPT*, 9 June 2000.

17. Kate Bramson, "Burrillville, R.I., Denies Renewal License for Junkyard," *Providence Journal*, 13 February 2003.

18. Phil Skinner, "Harmon Auto Wrecking," *C&P*, January 2001, 44–45, 47.

19. Tom Kertscher, "Oak Creek Considers Imposing $156,000 Fee for Junkyard Cleanup," *MJS*, 13 April 2009; Kertscher, "Oak Creek Likely to Impose Special Charge on Junkyard," *MJS*, 30 April 2009; Kertscher, "Oak Creek to Vote Tuesday on Billing Man for Removal of Junk," *MJS*, 22 May 2009.

20. "You Want It, You Got It," *American Bar Association Journal*, June 2002, 22.

21. Victoria Hurley-Schubert, "Spotswood, N.J., Officials, Car Company Reach Agreement over Junkyard," *Home News Tribune*, 24 October 2004.

22. Tu-Uyen Tran, "Grand Forks, N.D., to Discuss New Restrictions with Auto Junkyards," 3 June 2004 (quotes), and "Junkyard Permits Get Thumbs Up," 9 December 2005, both in *Grand Forks Herald*.

23. Some examples: Mary Beth Lane, "Neighbors Oppose Revival of Junkyard," *Columbus Dispatch*, 16 January 2007; Bob Stevens, "Revisiting Old Car City," *C&P*, January 1995, 24–26, 28; Sylvia Cooper, "Augusta, Ga., Recycler Slips Past New Ban," *Augusta Chronicle*, 8 June 2004; Tara Mack, "Junkyard Is a Site for Sore Feelings," *WP*, 14 September 1997.

24. Joe Heacock Jr., "Hap Gemmill's Junkyard," *C&P*, May 1990, 12–14; John Robertson, "Endangered Junkyard," *C&P*, September 1990, 51–54; Ron Kowalke, "Hauf Closing 63-Year-Old Oklahoma Salvage Business," *OCW*, 2 July 2009, 22.

25. Yards can operate profitably on small plots. See for example, Bob Stevens, "Papke Enterprises Might Just Be the World's Smallest Salvage Yard," *C&P*, April 1991, 58–60, 62; Mark J. Hash, "Oregon's Real McCoy," *C&P*, July 1992, 16–18; Jeff Babcock, "Revolvstore," *C&P*, April 1996, 15–17; Phil Skinner, "Renegade Wrecker, Law Abiding Citizen," *C&P*, December 1999, 56–57, 59; and Joe Sharretts, "Chick's Auto Parts Salvage Yard," *C&P*, February 2007, 64–68. But in general, acreage means inventory, and inventory means profitability.

26. John Pearley Huffman, "Midnight in the Garden of Eldorados and E-Types," *C&D*, September 2005, 112–16, 118; see also above, chap. 2.

27. Sheena Delazio, "Neighbors Upset over Property," *Time Leader* (Wilkes-Barre, Pa.), 28 December 2006. Old machinery and derelict cars on residential property are often associated with the poor, rural, white culture derogatively labeled "white trash." Consequently, homeowners often fear that their neighborhoods will become "white trash areas" if their neighbors collect them. On derelict cars and "white trash" culture, see John Hartigan Jr., "Unpopular Culture: The Case of 'White Trash,'" *Cultural Studies* 11 (1997): 318.

28. "Site Cleanup, Reservoir Level Discussed," *WTG*, 26 October 2007.

29. Bridget Hall, "Trash Is Man's Treasure, But County Wants It Gone," *SPT*, 11 March 2000. The Groff case was broadly similar.

30. Some examples: Lynne K. Varner, "Junkyard Stirs Ire of Neighbors," *WP*, 9 April 1992; Deanna Bellandi, "Holiday Tow Operator Has a Lot to Cope With," *SPT*, 13 April 1995; David T. Turcotte, "Gardner Panel Details Complaints on Junkyard," *WTG*, 2 November 1995; Karen Nugent, "Board Targets Town 'Junkyards,'" *WTG*, 9 January 1997; Zachary R. Mider, "West Warwick, R.I., Officials Again Seek Cure for Old Auto Shop, 'Eyesore,'" *Providence Journal*, 8 January 2004; Nate Hubbard, "Dispute Resolving," *Bland County Messenger* (Wytheville, Va.), 15 April 2009.

31. Eugene L. Meyer, "To Farmer Arnold, 'Automobile Junkyard' Is a Haymow," *WP*, 18 May 1982.

32. Bob Stevens, "Ohio Township Sells Yard Full of Cars," *C&P*, February 1999, 62–63; Bob Stevens, "Confiscated Collection Offered by Salvage Company," *C&P*, March 1999, 44–47.

33. David Cho, "Taking On a Backyard Battle," *WP*, 4 August 2002; Metro: Virginia, *WP*, 14 August 2002; "Fairfax County Wastes Tax Dollars Harassing Citizen," *WT*, 9 August 2002.

34. See, for example, Jane H. Lii, Neighborhood Report: Bayside, *NYT*, 9 September 1995; Phil Skinner, "Ford Fancier Tends to His Private Yard," *C&P*, April 1999, 47–49; Katherine Gazella, "Code Case Pits Tarpon Springs, Man's Livelihood," *SPT*, 10 December 2000; M. K. Guetersloh, "Local Woman Fined $1,430 for Leaving Many Vehicles on Property," *Pantograph* (Bloomington, Ind.), 14 May 2008; Chris Rickert, "Owner Sees a 'Few Cars'; Neighbor Sees a 'Junkyard,'" *Wisconsin State Journal*, 15 March 2009.

35. Carl Olson, SEMA Scene, *HRM*, November 1975, 13; SEMA Scene, *HRM*, December 1978, 2; Simone and Myers, "Weekend Warriors," esp. 136–37; George Wuerthner, ed., *Thrillcraft: The Environmental Consequences of Motorized Recreation* (White River Junction, Vt., 2007); David N. Lucsko, *The Business of Speed: The Hot Rod Industry in America, 1915–1990* (Baltimore, 2008), chaps. 7–9; "Legislation Introduced in Illinois to Ease Restriction on Antiques," *C&P*, May 1969, 47 (which, in spite of its title, explores how Illinois enthusiasts with old cars remained subject to various usage restrictions); Tool Bag, *C&P*, July 1969, 42.

36. Zimring, "Neon, Junk, and Ruined Landscape," esp. 99. In an exhaustive analysis of a variety of enthusiast titles, I found but one endorsement of the beautification project, as opposed to more than three hundred letters, editorials, and features voicing concern and outrage over the junkyard-clearing and anti-old-or-ratty-looking outlook that grew out of it (the exception: Sally Wimer and Gordon Jennings, "Lots of Luck Lady Bird," *C&D*, November 1966, 94–95).

37. Richard Sisson, Reader Forum, *C&P*, July 1990, 170; In the Headlights, *C&P*, April 1991, 189; Skinner, "Harmon Auto Wrecking."

38. Burnell D. Harty, classified, *C&P*, October 1969, 33; "Rick," classified, *HVW*, July 1991, 118; Frank G. Startare, Reader Forum, *C&P*, March 1992, 150.

39. Anonymous (name withheld by the editors), Reader Forum, *C&P*, February 2001, 8.

40. Eric Kaminsky, Voice of the Hobby, *C&P*, January 1998, 43; Bob Stevens, Voice of the Hobby, *C&P*, December 1998, 25; Steve McDonald, State Update, *SN*, April 2010, www.sema.org/sema-news/back-issues/archive (accessed 24 September 2012). Understandably, proposals like these were even more offensive to the enthusiast community when old-car hobbyists were targeted specifically; see Jeffrey S. Solochek, "Ordinance to Remove Junk Vehicles Passes," *SPT*, 16 April 2003.

41. See also In the Headlights, *C&P*, February 1990, 72; Jeff Smith, Starting Line, *HRM*, November 1990, 7; Startare, Reader Forum.

42. On 1920s regulations and Connecticut's landmark law: "Ways of Controlling Automobile 'Graveyards,'" *American City*, October 1929, 171; "Where Can Good Automobiles Go When They Die?" *Literary Digest*, 22 February 1930, 59–61. See also Bruce Baker, "Sturbridge Selectmen Want Junk Cars Removed," *WTG*, 12 April 1989; In the Headlights, *C&P*, February 1990, 72; Bob Stevens, Voice of the Hobby, *C&P*, February 1996, 73; Marlan Davis, Government Watch, *HRM*, August 2003, 13; and Evan Spring, Letter of the Month, *HRM*, July 2004, 15.

43. See Joe Mayall, Curbside, *SS*, October 1984, 4 (restoration-work ban); Patrick Bedard, "We've Seen the Ugly American, and We Don't Want It Parked Here," *C&D*, July 1983, 158 (truck ban); and below, chap. 6 (scrappage).

44. Bob Stevens, Voice of the Hobby, *C&P*, February 1996, 73.

45. Ken New, "Where Rust Stands Still!" *C&P*, July 1990, 40–43; Mike McCarthy, "Neighbors Collide with Sunrise Junkyard Plan," *Business Journal* (Sacramento, Calif.), 8 May 1995; Brent Hunsberger, "Profits in Parts," *Oregonian* (Portland), 6 February 2005; Joseph A. Slobodzian, "Junkyard Proving a Hard Sell in Phila.," *PI*, 31 August 2006.

46. Lucsko, *Business of Speed*, chaps. 7–9.

47. See below, chap. 6.

48. The SAN: Voice of the Hobby, *C&P*, January 1997, 50; "SEMA Action Network," *CHP*, January 2001, 19; "*Hot Rod* Power Tour, Day Five," *HRM*, June 2002, 80–81. ARMO: Steve Campbell, Starting Line, *HRM*, June 1993, 7; "A History of ARMO," *SN*, July 2011, 64.

49. "Cruise Night Sponsor," *HRM*, October 2002, 38–39; Marlan Davis, Government Watch, *HRM*, March 2003, 16; Steve McDonald, State Update, *SN*, May 2004, 111; Stephen Kim, "Is Your State Cool? We Take a Look at the Five Best States for Hot Rodders," *PHR*, May 2011, 42–44.

50. ARMO advertisements: *C&P*, December 1994, 63; January 1995, 45; April 1995, 129. SEMA Action Network advertisements: *EC*, April 2007, 103, and April 2009, 87; *Eurotuner,* July 2008, 89; *HRM*, April 2009, 109; *MT*, April 2007, 153, July 2008, 133, and April 2009, 130; *PHR*, April 2007, 45 (quote). ARMO also disseminated information on local issues to its members; see In the Headlights, *C&P*, February 2006, 14.

51. See, for example, Steve McDonald, State Legislation/Regulation, *SN*, December 1997, 10; McDonald, State Update, *SN*, August 2000, 10, and September 2005, 82–83; Brian Caudill, "SEMA Corner," *HRM*, July 2001, 22, and October 2001, 22; Bob Stevens, Voice of the Hobby, *C&P*, March 1996, 50, and May 1999, 22; and West Peterson, compiler, Voice of the Hobby," *C&P*, March 2001, 65.

52. On WOAH, COVA, the AACA, the CCCA, and others, see Bob Stevens, Voice of the Hobby, *C&P*, April 1993, 15, June 1993, 63, and July 1993, 71–72; and Len J. Athanasiades, Letter of the Month, *HRM*, September 1994, 8.

53. Bob Stevens, Voice of the Hobby, *C&P*, April 1993, 15, July 1993, 71–72, February 1996, 73 (quote), May 1996, 72, and July 1997, 55; Eric Kaminsky, Voice of the Hobby, *C&P*, November 1999, 60.

54. Steve McDonald, State Legislation/Regulation, *SN*, December 1997, 10; Eric Kaminsky, Voice of the Hobby, *C&P*, January 1998, 43, February 1998, 40, and May 1998, 40; Bob Stevens, Voice of the Hobby, *C&P*, December 1998, 25, January 1999, 26, and June 1999, 60; Steve McDonald, State Update, *SN*, June 2006, 64, August 2006, 62, February 2007, 209, and March 2007, 90; News and Notes, *C&P*, November 2006, 11.

55. West Virginia: West Peterson, Voice of the Hobby, *C&P*, June 2000, 29, and September 2000, 37; Steve McDonald, State Legislation/Regulation, *SN*, May 2000, 11; McDonald, State Update, *SN*, August 2000, 10, April 2006, 71, August 2006, 62, April 2007, 90, July 2007, 23, April 2008, 23, May 2009, 15, and June 2009, 21. Illinois: West Peterson, Voice of the Hobby, *C&P*, February 2000, 43; Driving Force, *S&CN*, January 2003, 29, 38; McDonald, State Update, *SN*, May 2004, 111, July 2004, 55, December 2004, 101, April 2005, 87, June 2005, 69, March 2006, 83, August 2006, 62, and June

2007, 83. Maine: McDonald, State Update, *SN*, August 2003, 88, and September 2005, 82–83. New Hampshire and Oregon: Bob Stevens, Voice of the Hobby, *C&P*, June 1999, 60; Brian Caudill, SEMA Corner, *HRM*, July 2001, 22; Fact and Rumor, *CC*, May 2002, 27; McDonald, State Update, *SN*, August 2002, 74; "Legislative 'Quick' Hits," *S&CN*, October 2002, 38.

56. Steve McDonald, State Update, *SN*, December 2004, 101, June 2005, 69, and July 2007, 23; Kim, "Is Your State Cool?"

57. Lynch, *Wasting Away*, esp. 1–10. Vance Packard describes an obsolescence-obsessed and diabolically wasteful dystopia, similar to Lynch's first, in *The Waste Makers* (New York, 1960), chap. 1. See also below, chap. 6.

58. David Edgerton's *Shock of the Old: Technology and Global History since 1900* (New York, 2007) emphasizes this point.

59. Automobility cannot function without junkyards. On the one hand, when de facto junkyards are cleared, legal junkyard businesses expand. On the other hand, when legitimate junkyards are zoned out of a given area (or when such businesses face sufficient licensing and regulatory hurdles that they all decide to close), de facto junkyards thrive, as happened on the Hawaiian island of Kauai in the late 1990s. See Jim Carlton, "Makeshift Junkyards Take Root on Kauai, and Scenery Suffers," *WSJ*, 19 February 1999.

6. Of Clunkers and Camaros

1. See, for example, *Cars and Parts'* "Peggy Sue" 1957 Chevrolet (March 1991–June 1993); *European Car's* 1951 Porsche 356 project (September 1993–October 1996); *Hot VWs'* "Project Streetwise" (August 1993–February 1995); and *Hot Rod's* "Project Cheap Thrills" Dart (March–May 1995).

2. "CAR" stood for "Chevron Auto Recycling."

3. See Christine Meisner Rosen, "Industrial Ecology and the Transformation of Corporate Environmental Management: A Business Historian's Perspective," in *Inventing for the Environment*, ed. Arthur Molella and Joyce Bedi (Cambridge, Mass., 2003), esp. 329–31.

4. David Freiburger, "Clunker Rescue," *HRM*, February 1994, 56–57; Rob Kinnan, "Crusher Camaro, Part 6," *HRM*, May 1995, 60–64, 66; "Where It All Began," *HRM*, December 2009, 22.

5. Unocal, "Corporate Responsibility," *Unocal 1990 Annual Report* (Los Angeles, 1990), 28–29.

6. Scrappage, therefore, was an example of "corporate greening" of the sort described in Finn Arne Jørgensen, *Making a Green Machine: The Infrastructure of Beverage Container Recycling* (New Brunswick, N.J., 2011), esp. 17 and 101: like beverage companies that supported bottle-deposit laws because they helped deflect the blame for roadside litter, Unocal and other corporate scrappage advocates saw the crusher as a way to appeal to regulators and "green consumers" and thus deflect attention from their problematic smokestacks.

7. Tom McCarthy, *Auto Mania: Cars, Consumers, and the Environment* (New Haven, Conn., 2007), 205.

8. On stakeholder framing as a stumbling block, see Barbara Gray, "Strong Opposition: Frame-Based Resistance to Collaboration," *Journal of Community and Applied Social Psychology* 14 (2004): 167 (quote); and "Framing of Environmental Disputes," in *Making Sense of Intractable Environmental Conflicts: Concepts and Cases*, ed. Roy J. Lewicki, Barbara Gray, and Michael Elliott (Washington, D.C., 2003), esp. 11–12.

9. Susan Strasser, *Waste and Want: A Social History of Trash* (New York, 1999); Carl Zimring, *Cash for Your Trash: Scrap Recycling in America* (New Brunswick, N.J., 2005); Zimring, "Neon, Junk, and Ruined Landscape: Competing Visions of America's Roadsides and the Highway Beautification Act of 1965," in *The World beyond the Windshield: Roads and Landscapes in the United States and Europe*, ed. Christof Mauch and Thomas Zeller (Athens, Ohio, 2008), 94–107; Zimring, "The Complex Environmental Legacy of the Automobile Shredder," *Technology and Culture* 52 (July 2011): 523–47; McCarthy, *Auto Mania*; Michael Thompson, *Rubbish Theory: The Creation and Destruction of Value* (New York, 1979); Kevin Lynch with Michael Southworth, *Wasting Away: An Exploration of Waste* (San Francisco, 1990); Shi-Ling Hsu and Daniel Sperling, "Uncertain Air Quality Impacts of Automobile Retirement Programs," *Transportation Research Record* 1444 (1994): 90–98; Robert W. Hahn, "An Economic Analysis of Scrappage," *RAND Journal of Economics* 26, no. 2 (Summer 1995): 222–42; Anna Alberini, Winston Harrington, and Virginia McConnell, "Estimating an Emissions Supply Function from Accelerated Vehicle Retirement Programs," *Review of Economics and Statistics* 78, no. 2 (May 1996): 251–65; Jennifer Lynn Dill, "Travel Behavior and Older Vehicles: Implications for Air Quality and Voluntary Accelerated Vehicle Retirement Programs" (Ph.D. diss., University of California, Berkeley, 2001); Dill, "Estimating Emissions Reductions from Accelerated Vehicle Retirement Programs," *Transportation Research, Part D* 9 (2004): 92; Hyung Chul Kim, "Shaping Sustainable Vehicle Fleet Conversion Policies Based on Life Cycle Optimization and Risk Analysis" (Ph.D. diss., University of Michigan, 2003), esp. 2–3, 170; Ryan Sandler, "Clunkers or Junkers? Adverse Selection in a Vehicle Retirement Program," working paper, 24 May 2011, "Miscellaneous" folder, SEMA-DC.

10. By the 1980s, "classic car" had assumed a new and much broader meaning. Although certain groups of gearheads would continue to use the term specifically to refer to high-end prewar cars, most enthusiasts—more to the point, most within the general public— came to use "classic car," more or less interchangeably with "vintage car," to refer to collectable vehicles of all ages. Because enthusiasts, public officials, and corporate spokespersons used "classic car" in this broader sense during their squabbles over accelerated vehicle retirement, I use the term that way in this chapter as well.

11. Thompson, *Rubbish Theory*, 6–8.

12. See James E. Krier and Edmund Ursin, *Pollution and Policy: A Case Essay on California and Federal Experience with Motor Vehicle Air Pollution, 1940–1975* (Berkeley, Calif., 1977); Robert W. Crandall et al., *Regulating the Automobile* (Washington, D.C., 1986), chap. 5; Gary C. Bryner, *Blue Skies, Green Politics: The Clean Air Act of 1990 and Its Implementation*, 2nd. ed. (Washington, D.C., 1995); Rudi Volti, "Reducing Automobile Emissions in Southern California: The Dance of Public Policies and Technological Fixes," in Molella and Bedi, *Inventing for the Environment*, 277–88; McCarthy, *Auto*

Mania, chaps. 5–10; and David N. Lucsko, *The Business of Speed: The Hot Rod Industry in America, 1915–1990* (Baltimore, 2008), chaps. 7–9.

13. McCarthy, *Auto Mania*, chaps. 8–10; Lucsko, *Business of Speed*, chap. 9; John Heitmann, *The Automobile and American Life* (Jefferson, N.C., 2009), 174. In *Saving the Planet: The American Response to the Environment in the Twentieth Century* (Chicago, 2000), 125, Hal K. Rothman emphasizes that air-pollution mitigation enjoyed broad bipartisan support in the 1960s–1980s. In the 1990s–2000s, bipartisan support among lawmakers for accelerated vehicle retirement was equally strong. Thus, although this chapter identifies less-well-known politicians' states (or districts) and party affiliations as a matter of convention, it warrants emphasis that East and West, North and South, and left and right, politicians flocked to the crusher as an environmental solution.

14. Dan Fisher, "Killing of Smog Bill Called 'Plain Crazy,'" *LAT*, 13 October 1971; Gladwin Hill, "Auto Smog Device Stirs California," *NYT*, 15 January 1974; Krier and Ursin, *Pollution and Policy*, 160–64, 240–47; McCarthy, *Auto Mania*, 204.

15. Initially, California inspections were primarily handled in roadside assessments or when a car was resold. Biannual tests for the majority of California vehicles began in 1984.

16. "Inside Detroit," *MT*, November 1968, 10–12; U.S. Congress, Office of Technology Assessment (OTA), *Retiring Old Cars: Programs to Save Gasoline and Reduce Emissions* (Washington, D.C., 1992), 15; Mark D. Warden, "The Collector Car and Pollution," white paper, 27 January 1993, "COVA Articles," SEMA-DC; "More Miles," *R&T*, March 1999, 19.

17. Chip Jacobs, "Scrap Fever," *Los Angeles Business Journal*, 13 August 1990, 4; OTA, *Retiring Old Cars*, 1; Sharon Begley, "Cold Cash for Old Clunkers," *Newsweek*, 6 April 1992, 61; Roberta Barth, "Florez Collides with Classic Car Buffs over Bill to Tag Them with Emissions Rules," *Political Pulse*, 11 April 2003 (quote); Crandall et al., *Regulating the Automobile*, 102.

18. Mike Lamm, "What You Can Do," *Special Interest Autos*, May–June 1971, 4 (both quotes).

19. Bryner, *Blue Skies*, 100–101.

20. Unocal News Release, 26 April 1990, rpt. in *CR-S*, 8 May 1990, 9631–32; Unocal advertisement, *LAT*, 27 April 1990; Larry B. Stammer, "Cash In a Clunker for Clean Air," *LAT*, 27 April 1990; Matthew L. Wald, "A Bid for Scrap, to Cut Pollution," *NYT*, 27 April 1990; Stephen Chapman, "Oil Firm Offers to Buy Old Cars," *WT*, 27 April 1990; Sen. Roth, "Removal of Older Cars from the Road," *CR-S*, 8 May 1990, 9631–33. Dolly Jørgensen describes a conceptually similar "win-win" in "An Oasis in a Watery Desert? Discourses on an Industrial Ecosystem in the Gulf of Mexico Rigs-to-Reefs Program," *History and Technology* 25 (December 2009): 345.

21. Unocal advertisement, *LAT*, 15 May 1990; Patrick Lee, "Unocal Car-Junking Bid Gets $700 Gift," *LAT*, 16 May 1990; In the Headlights, *C&P*, September 1990, 72, and October 1990, 72; Jacobs, "Scrap Fever"; Michael Lew, "Give Me Your Tired, Your Rusty . . . ," *NYT*, 6 October 1990. CARB was California's central smog authority, but there were a number of locally managed Air Quality Management Districts (AQMDs) and Air Pollution Control Districts (APCDs) across the state.

22. Chuck Coyne, "Call for Action," *CHP*, October 1991, 4; Richard J. Stegemeier, Chairman, President, and Chief Executive Officer, "To Our Stockholders," *Unocal 1990 Annual Report*, 2–3; Unocal, "Corporate Responsibility"; Dill, "Travel Behavior," 7–8; Jacobs, "Scrap Fever"; OTA, *Retiring Old Cars*, 1; "Richard J. Stegemeier, Unocal Corp.," *Directors and Boards*, Summer 1991, 56; "Working for Cleaner Air," *Unocal 1992 Annual Report* (Los Angeles, 1992), 20.

23. Bryner, *Blue Skies*, 100, 111–17; Lew, "Give Me Your Tired"; California Environmental Protection Agency, Air Resources Board, "Key Events in the History of Air Quality in California," www.arb.ca.gov/html/brochure/history.htm. Population figures from www.census.gov. Mileage data from Paul Sorensen, "Moving Los Angeles," *Access* 35 (Fall 2009): 17; and Sierra Business Council, "Vehicle Miles Traveled, 1980–2010" (chart), www.sbcouncil.org/Vehicle-Miles-Traveled (all sites accessed 3 May 2012).

24. Public Law 101-549—Nov. 15, 1990, Sec. 108, Miscellaneous Guidance, rpt. in *A Legislative History of the Clean Air Act* (Washington, D.C., 1993), ed. Environment and Natural Resources Policy Division of the Congressional Research Service of the Library of Congress, 1:479–80. See also letter from John Russell Deane III, Trainum, Snowdon, Hyland and Deane, to Charles Blum, SEMA President, 14 November 1991, "Clunker Legislation, 102nd Congress," SEMA-DC. On 1970s SIPs, see Bryner, *Blue Skies*, 100; and McCarthy, *Auto Mania*, 194.

25. Sen. Roth, "Removal of Certain Model Year Vehicles from Use," *CR-S*, 1 February 1990, 1064–65; Sens. Roth, Baucus, Symms, and Chafee, "Amendment No. 1406 to Amendment No. 1293," *CR-S*, 28 March 1990, 5597; Roth, "Removal of Older Cars"; Matthew L. Wald, "For Cleaner Air, Scrap Dirty Cars, U.S. Is Told," *NYT*, 24 October 1991; DRI/McGraw-Hill news release, 23 October 1991, "Aftermarket: 3/6/92 Letter to Associations," SEMA-DC.

26. William L. Schroer, "Accelerated Retirement—Draft for Public Comment," 2 October 1991, "Aftermarket: 3/6/92 Letter to Associations," SEMA-DC; EPA Office of Mobile Sources, "Accelerated Retirement of Vehicles," March 1992, http://epa.gov /orcdizux/stateresources/policy/transp/tcms/accelerate_retire.pdf (accessed 14 August 2009); Memorandum from John Russell Deane to Charles Blum, 23 July 1991, "Legislation: 102nd Congress," SEMA-DC; EPA Office of Mobile Sources, "Guidance for the Implementation of Accelerated Retirement of Vehicles Programs," February 1993, www.epa.gov/oms/stateresources/policy/scrapcrd.pdf (accessed 13 June 2011).

27. Bob Davis and John Harwood, "Bush Plans to Offer Pollution Credits to Firms That Buy, Then Junk Old Cars," *WSJ*, 9 March 1992; John E. Yang, "Bush Plans to Target Older Cars," *WP*, 10 March 1992. See also statement by President George H. W. Bush, 18 March 1992, White House Office of the Press Secretary, "Fact Sheet on Transportation Reforms," 18 March 1992, and Transcript, Press Briefing, 18 March 1992, all in "Bush Proposal," SEMA-DC.

28. In its MERC work, the EPA drew on an earlier report: "Emissions Trading Policy Statement," 4 December 1986, 51 *Federal Register* 43814–60.

29. California scrappage is considered separately below; on the others, see U.S. Generating Company, "Delaware Vehicle Retirement Program," n.d. [1992–1993], "Aftermarket Policy Statement: 9/27/93," SEMA-DC; Hahn, "Economic Analysis";

Dill, "Travel Behavior," 39–40; David M. Bearden, "A Clean Air Option," 16 September 1996, Congressional Research Service Reports, http://digital.library.unt.edu/govdocs /crs/permalink/meta-crs-287:1 (accessed 14 August 2009); Illinois EPA, "Pilot Project for Vehicle Scrapping in Illinois," May 1993, "Aftermarket Policy Statement: 9/27/93," SEMA-DC; Stevenson Swanson, "Your Clunker Might Actually Pay a Dividend," *CT*, 29 September 1992; "Sell the Old Beater and Clean the Air," *CT*, 13 October 1992; Andrew Maykuth, "If They Belch, Sputter or Wheeze, Sun Wants Them," *PI*, 2 September 1993; Francis J. Cebula, "The Sun Company Vehicle Retirement Pilot Program," 6 January 1994, "Illinois Proposal," SEMA-DC. The involvement of General Motors in Chicago was opportunistic, like Ford's involvement with SCRAP in Southern California: pollution reduction via a general reduction in the overall age of the fleet rather than new-car sales per se was the driving force behind the scrappage movement throughout the 1990s and 2000s. Not until the federal CARS program of 2009, discussed at length toward the end of this chapter, would direct stimulus for new-car sales become a primary scrappage aim in the United States.

30. EPA, "Economic Incentive Program Rules: Final Rule," 7 April 1994, 59 *Federal Register*, www.gpo.gov/fdsys/pkg/FR-1994-04-07/html/94-6828.htm (accessed 7 September 2011).

31. Pennsylvania: Walter Reed, Voice of the Hobby, *C&P*, August 1994, 73; Bob Stevens, Voice of the Hobby, *C&P*, July 1996, 144; Lee Ivanusic, Letter of the Month, *HRM*, October 1996, 8. New Jersey: letter from George T. Wiesner of Manuel Auto Parts, Inc., and the Automotive Recyclers Association to the New Jersey Department of Environmental Protection, n.d. [mid-1993], "Illinois Proposal," SEMA-DC; Bob Stevens, Voice of the Hobby, *C&P*, June 1996, 72; Steve McDonald, State Legislation/Regulation, *SN*, April 1998, 7; Eric Kaminsky, Voice of the Hobby, *C&P*, May 1998, 40. Virginia: D'Vera Cohn, "Can't Beat 'Em, Buy 'Em," *WP*, 9 November 1992; Ian Calkins, State Legislation/ Regulation, *SN*, June 1996, 8; Coalition for Auto Repair Equality, "It's a Unanimous Vote," *CARE Correspondent*, Summer 1997, www.careauto.org/correspondent/97/Summer /arto.htm (accessed 16 August 2011). Colorado: Chris Kersting, State Legislation/Regulation, *SN*, February 1995, 9; "State Eyes 'Clunker' Fund," *Rocky Mountain News* (Denver), 21 July 1992; Mark Harris, "Critics Spew Questions at 'Clunkers' Program," *Rocky Mountain News* (Denver), 19 February 1993; "Old-Car Buy-Back Programs Are Real Clunkers," *OCW*, 22 February 1996. Florida: Ian Calkins, State Legislation/Regulation, *SN*, October 1995, 8–9; Steve McDonald, State Legislation/Regulation, *SN*, February 1999, 11; Bob Stevens, Voice of the Hobby, *C&P*, March 1999, 36. Louisiana: Ian Calkins, State Legislation/Regulation, *SN*, October 1995, 8–9; "Old-Car Buy-Back Programs Are Real Clunkers." Massachusetts: Chris Kersting, Federal Legislation/Regulation, *SN*, February 1995, 9. Michigan: Steve McDonald, State Legislation/Regulation, *SN*, January 1999, 13. Arizona: Ian Calkins, State Legislation/Regulation, *SN*, May 1996, 11; Calkins, State Legislation/Regulation, *SN*, February 1997, 7; Calkins, State Legislation/Regulation, *SN*, April 1997, 6; Eric Kaminsky, Voice of the Hobby, *C&P*, June 1998, 50; Steve McDonald, State Legislation/Regulation, *SN*, August 1998, 6. Vermont: Steve McDonald, State Legislation/Regulation, *SN*, June 1999, 13; Driving Force, *S&CN*, January 2003, 29, 38; Steve McDonald, Law and Order—State Update,

SN, September 2008, 21. North Carolina: Law and Order—State Update, *SN*, July 2009, 17; Law and Order—State Update, *SN*, September 2009, n.p. Washington: Richard Roesler, "Plan Would Trade Tax Break for Old Cars," *Spokesman-Review* (Spokane, Wash.), 5 January 2006.

32. Maine: Steve McDonald, State Update, *SN*, July 2000, 14; Dieter Bradbury, "State Postpones Buyback Program for Junk Vehicles," *Portland Press Herald*, 22 December 2000; Joseph McCann, "Maine Amends Payments for Scrapping Old Vehicles," *American Metal Market*, 28 June 2001, 6; McDonald, State Update, *SN*, March 2001, 6, and September 2001, 6. Illinois: McDonald, State Legislation/Regulation, *SN*, November 1999, 24–25; McDonald, State Update, *SN*, September 2000, 10; West Peterson, Voice of the Hobby, *C&P*, November 2000, 34. Texas: Fax from B. Carraghan, Automotive Service Industry Association, to Chris Kersting, Trainum, Snowdon, Hyland and Deane, 16 June 1993, "Mobile-Stationary Source Credit Trading Program," SEMA-DC; EPA, 40 CFR 52, 19 December 1997, 62 *Federal Register* 66576–78; Chris Kersting, Federal Legislation/Regulation, *SN*, March 1998, 6; Eric Kaminsky, Voice of the Hobby, *C&P*, May 1998, 40, and November 1999, 60; "Texas Scrappage Bill Would Overturn Repeal," *DF*, April 1999, 1; McDonald, State Legislation/Regulation, *SN*, May 1999, 16–17, August 1999, 14, and February 2000, 13; "Texas Hobbyists Hang Tough," *DF*, July 1999, 1; "Texas Proposes Scrappage Program," *DF*, February 2000, 1; Peterson, Voice of the Hobby, *C&P*, March 2000, 39, and November 2000, 34; Texas Natural Resource Conservation Commission, "Chapter 114—Control of Air Pollution from Motor Vehicles, Subchapter F," 19 April 2000, "Texas Proposal," SEMA-DC; McDonald, State Update, *SN*, September 2000, 10, and September 2001, 6; "Urgent Legislative Alert," 28 June 2001, "Texas Proposal," SEMA-DC; Gordon Dickson, "AirCheck Program Helps Repair over 10,000 Cars," *Fort Worth Star-Telegram*, 15 May 2006; Kim, "Is Your State Cool?" Chisum, a Democrat when first elected, became a Republican in 1996.

33. "Marketable Permit Program under Study," *Advisor*, Spring 1991, 1–2.

34. Unocal news release, 26 May 1993, unarchived CARB records, Sacramento (U-ARB); Michael Parrish, "Unocal Seeks 500 Cars for Clean-Air Tests," *LAT*, 27 May 1993; CARB news release, 26 May 1993, U-ARB. At the time this research was conducted, there was a gap in publicly accessible CARB records. Those into the early 1980s were at the state archives, and those since 1996 were at www.arb.ca.gov, but those from the early 1980s through 1995 remained in limbo at CARB's offices. This included a number of critical documents; I thank Alexa Barron of CARB's legal affairs department for locating these records and making them available.

35. Letter from Mike Reihle, Unocal, to Raphael Susnowitz, Mobile Source Division, CARB, 28 October 1992; CARB, "Co-Funding Documentation, Contract No. AB2766/C930xx," n.d. [1992–1993]; letter from John Prudent, General Manager, Hugo Neu-Proler, to Bob Kniesel, Intergovernmental Affairs Office, SCAQMD, 20 July 1992; letter from Tom Bradley, mayor of Los Angeles, to Richard J. Stegemeier, 24 July 1992; letter from Diane Ward, Field Operations Division, State of California Department of Motor Vehicles, to Bob Kniesel, 23 July 1992; letter from K. D. Drachand, Chief, Mobile Source Division, CARB, to Bob Kniesel, 21 July 1992 (all in U-ARB).

36. "Want to Pollute? Get a Clunker off the Street," *San Francisco Examiner*, 10 January 1993; "RECLAIM," *Advisor*, Summer 1992, 1; "Working for Cleaner Air"; "Creative Methods to Help Clean the Air," *Unocal 1993 Annual Report* (Los Angeles, 1993), 20; "500 Old Cars to Be Scrapped in Smog Trade-Off Plan," *LAT*, 13 May 1993.

37. "SCRAP II Press Conference: Remarks by Richard J. Stegemeier," 26 May 1993, U-ARB; Unocal news release, 26 May 1993, U-ARB; "SCRAP Saves Unocal $5 Million," *Unocal 1993 Annual Report*, 16.

38. My calculations, based on a comparison of SCRAP II data with the California standards applicable to model year 1994 cars (the newest available in 1993) and to 1970s models. CARB tested sixty vehicles during SCRAP II; its results appear in an untitled set of tables, U-ARB. California standards for 1994: Table 4.2, in *Structured Catalysts and Reactors*, 2nd ed., ed. Andrzej Cybulski and Jacob A. Moulijn (Boca Raton, Fla., 2006), 113. Standards for 1970s vehicles: EPA, "Summary of Current and Historical Light-Duty Vehicle Emission Standards," April 2007, www.epa.gov/greenvehicles/detailedchart.pdf (accessed 18 August 2011).

39. Richard J. Stegemeier, "Clunkers," *LAT*, 28 March 1992.

40. "The Aftermarket Response to the 'Clunker' Issue," 27 September 1993, "Aftermarket Policy Statement: 9/27/93," SEMA-DC; Oscar Suris, "Dirty Driving," *WSJ*, 17 August 1994; Eric Brockman, Voice of the Hobby, *C&P*, June 1994, 73; Gerald Perschbacher, "Unocal Forms Scrappage Subsidiary," *OCW*, 14 March 1996; Bob Stevens, Voice of the Hobby, *C&P*, May 1996, 72; Sam Atwood, "Monies to Scrap Old Cars Ok'd," *AQMD-A*, January 1996, 7.

41. Rule 1623: Sam Atwood, "Incentive to Scrap Old, Dirty Lawnmowers Adopted," *AQMD-A*, July 1996, 1, 3. The lawn-mower program: "Lawn Mower Buyback Program a Sellout," *AQMD-A*, July 2003, 3; "Lawn Mower Exchange Mows Down Pollution," *AQMD-A*, July 2006, 5; and "More than 9,000 Lawn Mowers Will Be Available for 2010 Exchange Program," *SCAQMD Advisor*, March 2010, 2. The leaf-blower program: "Leaf Blower Exchange for Landscape Maintenance Professionals," *AQMD-A*, March 2006, 5; "Board Approves a Leaf Blower Exchange for Landscape Maintenance Professionals," *AQMD-A*, March 2007, 5; and "Registration for Low-Polluting Leaf Blower Exchange Begins July 15, 2008," *SCAQMD Advisor*, July 2008, 6.

42. Sam Atwood, "Board Overhauls Rideshare Rule," *AQMD-A*, May 1995, 1; Bill Kelly, "February Rule Actions," *AQMD-A*, March 1999, 2; "Officials Replace Controversial Rideshare Rule," *AQMD-A*, January 1996, 1, 3.

43. Ventura County: Tina Daunt, "Car-Pool Alternative," *LAT*, 5 May 1993; Daunt, "Anti-Smog Business Fees to Be Voted On," *LAT*, 24 August 1993. Sacramento: Bob Stevens, Voice of the Hobby, *C&P*, February 1994, 74. Bay Area: Bob Stevens, Voice of the Hobby, *C&P*, March 1996, 50; Jennifer Dill, "Scrapping Old Cars," *Access*, Spring 2004, 22–27; Sandler, "Clunkers or Junkers?" Other AQMDs and APCDs: Dill, "Travel Behavior," 39–40; CARB, "Report to the California Legislature: Accelerated Light-Duty Vehicle Retirement Program," July 2004, www.arb.ca.gov/msprog/avrp/FINAL_Legislative_Scrap_Report07-04.pdf (accessed 16 March 2011), 5.

44. Dill, "Travel Behavior," 39–40; CARB, "Staff Report," 30 November 2001, 3, "Calif.: State Legislation: S. 501: Implementing Regs.: 2002 & SCAQMD," SEMA-DC; CARB, "Report to the Legislature," 6–7.

45. CARB, "Report to the Legislature," 2; Kersting, Federal Legislation/Regulation, *SN*, February 1995, 9.

46. Letter from Robert F. Hellmuth, SEMA, to Senator Charles Calderon, 7 July 1995, "Calif.: State Legislation: S. 501 (1995)," SEMA-DC; Kersting, Federal Legislation/Regulation, *SN*, November 1995, 6–7; memorandum from Anna Phillips to Bob Hellmuth and Chuck Blum of SEMA, 28 August 1995, "Calif.: State Legislation: S. 501 (1995)," SEMA-DC.

47. CARB, "Guidelines for the Generation and Use of Mobile Source Emission Reduction Credits," April 1993, "Calif.: ARB: Clunker Guidelines: Mobile Source Credits: Final Guidelines (4/93)," SEMA-DC.

48. K. D. Drachand, Chief, Mobile Source Division, CARB, "Notice of Workshop Concerning Development of Voluntary Light-Duty Vehicle Accelerated Retirement Program Regulations," 9 February 1996, "Calif.: State Legislation: S. 501: Implementing Regs.: 1996," SEMA-DC; CARB, "Notice of Workshop Concerning Development of Voluntary Accelerated Vehicle Retirement Program Regulations," 4 February 1997, www.arb.ca.gov/msprog/avrp/vavrwshp.pdf (accessed 16 March 2011); CARB, "Staff Report—Initial Statement of Reasons," 23 October 1998, www.arb.ca.gov/regact/scrap/isor.pdf (accessed 16 March 2011); CARB, "ARB Resolution 98-64," 10 December 1998, www.arb.ca.gov/regact/scrap/res98-64.pdf (accessed 16 March 2011); CARB, "Regulations for Voluntary Accelerated Light-Duty Vehicle Retirement Enterprises," 22 October 1999, www.arb.ca.gov/regact/scrap/finreg.pdf (accessed 16 March 2011).

49. CARB, "Staff Proposal: Proposed Amendments to Air Resources Board Voluntary Accelerated Vehicle Retirement Regulations," 9 October 2001, www.arb.ca.gov/msprog/avrp/100901staffprop.pdf (accessed 16 March 2011); CARB, "Staff Report: Resolution of Differences Between BAR and ARB Voluntary Accelerated Vehicle Retirement Regulations and Options to Address Parts Recovery and Resale from Retired Vehicles," 25 June 2001, "Calif.: State Legislation: S. 501: Implementing Regs.: 2001," SEMA-DC; CARB, "Staff Report—Initial Statement of Reasons for Proposed Rulemaking," 30 November 2001, "Calif.: State Legislation: S. 501: Implementing Regs.: 2002 & SCAQMD," SEMA-DC.

50. CARB, "Notice of Public Workshop to Discuss Revisions to the Voluntary Accelerated Vehicle Retirement Regulation," 30 May 2006, www.arb.ca.gov/msprog/avrp/WorkshopNoticeMSC06-09.pdf (accessed 16 March 2011); CARB, "Notice of Public Workshop to Discuss Revisions to the Voluntary Accelerated Vehicle Retirement Regulation and Carl Moyer Program Guidelines for Voluntary Repair of Vehicles," 3 August 2006, www.arb.ca.gov/msprog/avrp/MSC06-11VAVR_WkshpNtc.pdf (accessed 16 March 2011); CARB, "Final Statement of Reasons for Rulemaking," 7 December 2006, www.arb.ca.gov/regact/vavr06/fsor.pdf (accessed 16 March 2011).

51. CARB, "Report to the California Legislature," A-2. The pilot program was conducted in Southern California (Dill, "Estimating Emissions Reductions").

52. "Urgent Regulatory Alert," draft, n.d. [May or early June 2009], "Miscellaneous," SEMA-DC; CARB, "Final Statement of Reasons for Rulemaking," 26 June 2009, "Calif.: State Legislation: S. 501: Implementing Regs.: 2005, 2006, 2007, 2008, 2009–," SEMA-DC.

53. Letter from George T. Wiesner to the New Jersey Department of Environmental Protection, n.d. [mid-1993], "Illinois Proposal," SEMA-DC.

54. CARB, "Staff Report," 23 October 1998, 16 (emphasis in original); CARB, "Regulations for Voluntary Accelerated Light-Duty Vehicle Retirement Enterprises."

55. Hsu and Sperling, "Uncertain Air Quality Impacts"; Hahn, "Economic Analysis"; Alberini, Harrington, and McConnell, "Estimating an Emissions Supply Function"; Dill, "Estimating Emissions Reductions"; Sandler, "Clunkers or Junkers?"

56. Letter from Mark E. Simons, Emission Control Strategies Branch, EPA, to Anna Phillips, SEMA, 9 March 1994, "Calif.: SCAQMD: Rule 1610: Proposed Amended Rule," SEMA-DC; see also CARB, "Staff Report," 23 October 1998, 16.

57. Lew, "Give Me Your Tired"; California EPA, ARB, "Key Events in the History of Air Quality"; U.S. Department of Transportation, Bureau of Transportation Statistics, Table 1-11, www.bts.gov/publications/national_transportation_statistics/html/table_01_11.html (accessed 4 May 2012).

58. Thomas D. Larson, Federal Highway Administrator, *Highway Statistics 1990* (Washington, D.C., 1990), 17. California's climate exacerbated its smog problems; see McCarthy, *Auto Mania*, 116, 118–19.

59. Hahn, "Economic Analysis"; Anna Alberini, Winston Harrington, and Virginia McConnell, "Fleet Turnover and Old Car Scrap Policies," March 1998, www.rff.org/documents/RFF-DP-98-23.pdf (accessed 14 August 2009); Sandler, "Clunkers or Junkers?"; Steve McDonald, State Update, *SN*, September 2001, 6.

60. Initially, the program targeted pre-1987 cars, but for the reasons cited in the previous paragraph the date was subsequently shifted to 1996, then 1998.

61. Bradbury, "State Postpones Buyback"; McCann, "Maine Amends Payments."

62. OTA, *Retiring Old Cars*; Hsu and Sperling, "Uncertain Air Quality Impacts."

63. Comments of Sierra Club director Daniel Becker, in Rudy Abramson, "Administration OKs 'Cash for Clunkers' Plan," *LAT*, 19 March 1992; Paul Rauber, "Schemes That Go Clunk," *Sierra*, July–August 1992, 38; Bob Golfen, "'Cash for Clunkers' Is Nonsense," *Arizona Republic* (Phoenix), 3 April 1993.

64. Golfen, "Nonsense"; Maria L. La Ganga, "Anti-Smog Program Offers Hazy Idea of Older Cars," *LAT*, 2 May 1993; "Want to Pollute?"; Harris, "Critics Spew Questions."

65. Robert Stevens, "Civil Rights Laws Are Cited in Challenge to California Pollution-Control Tactics," *WSJ*, 24 July 1997; "California Pollution Officials Suspend Emissions Trading for New Programs," *WSJ*, 29 August 1997; "Chairman Orders Committee to Review Auto Scrapping Credit Programs," *AQMD-A*, March 1998, 1; Eric Kaminsky, Voice of the Hobby, *C&P*, December 1997, 64. See also Rosen, "Industrial Ecology."

66. Lucsko, *Business of Speed*, chaps. 7–9 (quote on 165).

67. Jim Motavalli, "One Man's Junk," *E: The Environmental Magazine*, October 1994, 19.

68. Bobbie Weatherbee, "Car Buffs, Suppliers Losers on This Bill," *Wichita Eagle*, 6 October 1992; Robert Kaiser, "'Clunker Bills' Threaten to Put the Old Ford to Sleep," *Buffalo News*, 29 March 1993; M. B. Rebedeau, "Those 'Clunkers' May Have Other Value," *CT*, 8 November 1992; Alexander Cockburn, "A Kind, Clean Word for the Clunker," *LAT*, 15 March 1992; Stegemeier, "Clunkers"; "Let True Classics Enjoy Old Age" (three letters), *WSJ*, 16 September 1994.

69. Golfen, "Nonsense"; La Ganga, "Anti-Smog Program Offers Hazy Idea of Older Cars"; "Want to Pollute?"

70. Pat Brennan, "A Classic Crush," *Orange County Register*, 26 September 1993 (quote). See also Motavalli, "One Man's Junk"; and Doug Smith, "Collectors Combat 'Clunker Bills,'" *Charlotte Observer*, 2 April 1993.

71. Letters: Reaction Time, *HRM*, July 1991, 8, 10; Reader Forum, *C&P*, March 1992, 151, and March 1993, 159; Melvin Pfaft, "Help for the 'Clunkers,'" *VWT*, October 1994, 11; Robert Walker, Letter of the Month, *HRM*, November 1993, 10. Editorials: Jim McGowan, Point of View, *CC*, October 1990, 5; Jeff Smith, Starting Line, *HRM*, October 1991, 7; Coyne, "Call for Action"; Wade Hoyt, Editor's Report, *Motor*, January 1994, 5, 27; "Clunkers," *V8 Power*, Spring 1994, 68–70; "12th Annual Vintage Special," *HVW*, July 1994, 46; Freiburger, "Clunker Rescue"; Freiburger, "Clunker-Law Tirade," *4-Wheel and Off-Road*, February 1996, 6.

72. Motavalli, "One Man's Junk" (quote); Bob Stevens, Voice of the Hobby, *C&P*, January 1994, 76–77.

73. On parts interchange, see "Junkyard Jamboree," *HRM*, February 1977, 70–74, 76–83; Cam Benty, "Salvage Yard Maze," *PHR*, August 1982, 76–79; and Terry McGean, "Getting Away with It," *HRM*, January 1999, 43.

74. Thompson, *Rubbish Theory*, esp. 6–7, 9–10. Strasser elaborates on the state of limbo in *Waste and Want*, 6–7.

75. Thompson, *Rubbish Theory*, 27.

76. Robert J. Gottlieb, Classic Comments, *MT*, February 1963, 78–80; Dominick D. Amato, Tool Bag, *C&P*, April 1974, 7-H; Paul Maddams, Letter of the Month, *HRM*, March 1996, 8; Tom Igoe, Letter of the Month, *HRM*, September 2004, 16; David Freiburger, Rob Kinnan, and Mike Finnegan, interview by author, El Segundo, California, 24 May 2011.

77. Richard Roesler, "Plan Would Trade Tax Break for Old Cars," *Spokesman-Review*, 5 January 2006; Jeff Koch, "Mope-A-Dope," *HRM*, October 1997, 74–76; Steve Magnante, "Bottle Bomb," *HRM*, February 1998, 42–47.

78. Joseph P. Smith, Reader Forum, *C&P*, July 1993, 159; William H. Bradshaw, Reaction Time, *HRM*, September 1995, 8.

79. McGowan, Point of View.

80. Lucsko, *Business of Speed*, chaps. 7–9.

81. Ibid., 195–200.

82. Rob Kinnan, "Crusher Camaro, Part 1," *HRM*, December 1994, 72–74; Kinnan, "Crusher Camaro, Part 3"; Freiburger, Kinnan, and Finnegan interview. See also Freiburger, "Clunker Rescue"; Kinnan, "Crusher Camaro, Part 6"; and "Where It All Began." The National Muscle Car Association sponsored a similar project, a 1969 Mustang from Project CAR (Headlights, *C&P*, August 1994, 75).

83. Jennifer Dill, "Design and Administration of Accelerated Vehicle Retirement Programs in North America and Abroad," *Transportation Research Record* 1750 (2001): 26; letter to Philip A. Lorang, EPA, 27 September 1993, "Aftermarket Letter to EPA: 9/27/93," SEMA-DC; Les Jackson, "Take Notice of Another 'Scrap Bill,'" *WT*, 8 February 2002; Brian Caudill, memorandum, 19 April 2002, "LEG: 107th Congress: Scrappage

Program within Comprehensive Energy Bill: Industry Letter to Senate & Action Alert to Members," SEMA-DC.

84. Charles P. Schwartz Jr., "Voice of the People," *CT*, 25 October 1992; J. Hansen, Reader Forum, *C&P*, November 1993, 164; Freiburger, "Clunker-Law Tirade"; Freiburger, Starting Line, *HRM*, August 2006, 18. See also Kim, "Vehicle Fleet Conversion Policies."

85. Letter to Lorang; draft letter to Mark Simons, EPA, 18 October 1993, "Ann Arbor Meeting: 10/6/93," SEMA-DC; SEMA, "Urgent Regulatory Action Alert," n.d. [fall 1998], "Calif.: State Legislation: S. 501: Implementing Regs.: 1998," SEMA-DC; John H. White, "Collectors of Old Cars Battle Clunker Laws," *Argus* (Fremont, Calif.), 24 May 1996.

86. La Ganga, "Anti-Smog Program"; Bob Stevens, Voice of the Hobby, *C&P*, August 1993, 73.

87. EPA Office of Mobile Sources, "Accelerated Retirement of Vehicles," 4.

88. See, for example, John McElroy, "Scrap the Clunkers," *AI*, February 1992, 5; and Katherine Shaver, "Older Cars Fouling Region's Air Quality," *WP*, 4 July 2009.

89. Walter Reed, Voice of the Hobby, *C&P*, August 1994, 73.

90. Steve Campbell, Starting Line, *HRM*, February 1994, 6; Maddams, Letter of the Month; Bill Barrile, "Old Car Lovers Fighting for a Say on 'Clunkers,'" *Buffalo News*, 11 April 1993; Arch Brown, Voice of the Hobby, *C&P*, August 1995, 74–75; Eric Brockman, Voice of the Hobby, *C&P*, June 1994, 73; "Texas Scrappage Bill Would Overturn Repeal"; Bob Stevens, Voice of the Hobby, *C&P*, May 1993, 73.

91. James J. Baxter, "A Bad Idea Whose Time Should Not Come," unpublished draft, 16 March 1992, "Press Clips: 1992," SEMA-DC; Bob Stevens, Voice of the Hobby, *C&P*, April 1993, 15; Stevens, Voice of the Hobby, *C&P*, June 1993, 63; Stevens, Voice of the Hobby, *C&P*, January 1994, 76–77.

92. Mike Magda, Skid Marks, *CHP*, April 1993, 4; Len J. Athanasiades, Letter of the Month, *HRM*, September 1994, 8; John R. White, "*Hemmings* Offers Seed Money to Help Keep the Old-Car Hobby in Bloom," *Boston Globe*, 4 September 1993.

93. Letter to Terri Wilsie, EPA, 14 June 1993, "EPA Program: Guidance Documents Comments: 6/14/93," SEMA-DC; letter to U.S. Rep. Sue W. Kelly, 1 August 1995, "LEG: 104th Congress: CMAQ: Highway Bill (S440/HR2274)," SEMA-DC; letter to U.S. Senator James M. Inhofe, 2 April 2003, "LEG: 108th Congress: CMAQ: ISTEA: Correspondence to Senate/House," SEMA-DC.

94. Letter from Len Athanasiades of Year One, Inc., to Don Turney, Charles Blum, George Elliot, and Russ Deane of SEMA, 17 April 1992, "Year One Inc.," SEMA-DC; "SREA Position Paper on the Clunker Issue," n.d. [early 1992], "Press Clips: 1992," SEMA-DC; Baxter, "A Bad Idea"; SEMA, "Proposed Aftermarket Program to Address State-Level Vehicle Scrappage Programs," n.d. [mid-1993], "Miscellaneous," SEMA-DC; News Briefs, *HRM*, April 1993, 17; ARMO advertisement, *C&P*, January 1995, 45, and *PHR*, January 1995, 95; Dick Wells, SEMA Public Relations, "Special Memo to Automotive Media: Governmental 'Clunker' Activities," 5 March 1993, "Sample Press Kit Materials," SEMA-DC.

95. Letter from Len Athanasiades to Don Turney, Charles Blum, George Elliot, and Russ Deane; SEMA, "Proposed Aftermarket Program"; Magda, "Skid Marks"; Voice of

the Hobby, *C&P*, January 1997, 50; Chris Kersting, interview by author, Diamond Bar, California, 26 May 2011; Steve McDonald and Stuart Gosswein, interview by author, Washington, D.C., 28 June 2011.

96. Memorandum from Frank Bohanan to Russ Deane, 20 August 1992, "Vehicle Emissions Data: 1992 SEMA Analysis," SEMA-DC (emphasis in original); "Testimony of Frank Bohanan on Behalf of the Specialty Equipment Market Association (SEMA), Presented to the California I/M Review Committee," 27 January 1993, "Aftermarket Policy Statement: 9/27/93," SEMA-DC.

97. Rae Tyson, "Air-Fouling San Diego Cars Get Free Fix-Up," *USA Today*, 8 August 1996; Dale Jewett, "Old Cars Can Run Cleaner," *AN*, 4 March 1996; Ian Calkins, State Legislation/Regulation, *SN*, October 1996, 6–7; SEMA, "Voluntary Repair and Upgrade as an Alternative to Motor Vehicle Scrappage Programs," white paper, 2004 (I thank SEMA's Peter MacGillivray for providing a copy of this document).

98. The list: "CARB Resolution 93-9: Collector Vehicle Exemption Guidelines (1955–1981)," n.d. [early 1993], "Calif.: ARB: Clunker Guidelines: Mobile Source Credits: Final Guidelines (4/93)," SEMA-DC. Its opposition: Bob Stevens, Voice of the Hobby, *C&P*, November 1993, 78–79; Reader Forum, *C&P*, February 1994, 159.

99. Michael J. Conlon, "Comments of the Motor Vehicle Aftermarket Industry on the Illinois Environmental Protection Agency's Proposed Vehicle Scrappage Program Rules," 14 December 1994, "Illinois Proposal," SEMA-DC; Steve McDonald, State Update, *SN*, December 2000, 12; SEMA, "Proposed Aftermarket Program"; Ian Calkins, "Statement of Specialty Equipment Market Association (SEMA) to California Air Resources Board Regarding Light-Duty Accelerated Vehicle Retirement and Repair Program, March 7, 1996," draft, "Calif.: State Legislation: S. 501: Implementing Regs.: 1996," SEMA-DC; John O'Dell, "Parts Seekers Get Help in Beating the Crunch," *LAT*, 8 May 2002; SEMA press release, 30 April 2002, "Miscellaneous," SEMA-DC.

100. SEMA, "Urgent Regulatory Action Alert," n.d. [Fall 1998], "Calif.: State Legislation: S. 501: Implementing Regs.: 1998," SEMA-DC; letter from Calif. State Senator Maurice Johannessen to Dr. William A. Burke of CARB, 20 December 2001, "Miscellaneous," SEMA-DC; Eric Kaminsky, Voice of the Hobby, *C&P*, November 1999, 60; CARB, "Staff Report," 30 November 2001.

101. Ian Calkins, State Legislation/Regulation, *SN*, February 1997, 7, and April 1997, 6; Bob Stevens, Voice of the Hobby, *C&P*, July 1996, 144; Walter Reed, Voice of the Hobby, *C&P*, June 1997, 41; Steve McDonald, State Legislation/Regulation, *SN*, April 1998, 6–7, and August 1998, 6; Eric Kaminsky, Voice of the Hobby, *C&P*, June 1998, 50.

102. See above, chap. 6, note 31.

103. Chris Kersting, Federal Legislation/Regulation, *SN*, October 1995, 8, and January 1996, 7; Council of Vehicle Associations, "COVALERT," n.d. [Fall 1995], "LEG: 104th Congress: CMAQ: Highway Bill (S440/HR2274)," SEMA-DC; memorandum from Stuart Gosswein, 27 September 1996, "LEG: 105th Congress: CMAQ: ISTEA: Aftermarket Letter," SEMA-DC; Steve McDonald, State Legislation/Regulation, *SN*, November 1997, 30; "This Is Serious," *HRM*, March 2002, 18; Brian Caudill, "Washington Report, *SN*, May 2002, 34; Automotive Service Association news release, 20 February 2002, "LEG: 107th Congress: Scrappage Program within Comprehensive Energy Bill:

Legislative Text: Scrappage Program," SEMA-DC; "Amendment No. 3007 to Amendment No. 2917 to S. 517," *CR-S*, 13 March 2002, S1838–46; SEMA press release, 3 August 2005, "LEG: 109th Congress: CMAQ: ISTEA: TEA 21: 2005 Reauthorization," SEMA-DC.

104. Alan S. Blinder, "A Modest Proposal," *NYT*, 27 July 2008; European Federation for Transport and Environment, "EU Should Not Encourage Subsidies for New Cars," 3 March 2009, "LEG: 111th Congress: Europe," SEMA-DC; George Monbiot, "This Scam Is Nothing But a Handout for Motor Companies, Resprayed Green," *Guardian* (U.K.), 24 March 2009; Rep. Miller, "'CARS' Bill a Prescription for Improved Auto Sales," *CR-H*, 31 March 2009, H4131–32; Nelson D. Schwartz, with Maria de la Blume, "'Euros for Clunkers' Drives Sales," *NYT*, 1 April 2009; Sens. Stabenow and Brownback, untitled, *CR-S*, 2 June 2009, S5933–36; Sen. Feinstein, untitled, *CR-S*, 8 June 2009, S6260–65; Sen. Casey, untitled, *CR-S*, 11 June 2009, S6553–56; Justin Hyde, "Cash-for-Clunkers Plan Runs into Competing Legislation," *Detroit Free Press*, 2 April 2009; Bob Gritzinger and Greg Kable, "Cash for Clunkers," *Autoweek*, 6 April 2009, 11; Federal Update, *SN*, April 2009, 22, and August 2009, 18; McDonald and Gosswein interview. Conceptually, the use of government largesse to spur new-car sales was an old idea. In 1931, for example, automaker André Citroën argued in favor of progressively punitive taxation on vehicles older than seven years in order to encourage their disposal and replacement with newer models: see John Reynolds, *André Citroën: The Man and the Motor Cars* (Stroud, Gloucestershire, 1996), 150. The use of *private* largesse toward the same end is an idea of roughly the same vintage: see above, chap. 1, and Steven M. Gelber, *Horse Trading in the Age of Cars: Men in the Marketplace* (Baltimore, 2008), esp. 71–72.

105. Ralph Vartabedian and Ken Bensinger, "Why 'Clunkers' Excludes Old, Polluting Cars," *LAT*, 13 August 2009; McDonald and Gosswein interview.

106. CARS success: Rep. Sutton, untitled, *CR-H*, 16 September 2009, H9573–74. Post-CARS slump: Rep. Brady, untitled, *CR-H*, 29 October 2009, H12128; John K. Teahen Jr., "After Clunkers, Sales Crash," *AN*, 5 October 2009, and Resources for the Future press release, 4 August 2010, both in "LEG: 111th Congress: Cash for Clunkers—Post Program," SEMA-DC; McDonald and Gosswein interview.

107. Vance Packard, *The Waste Makers* (New York, 1960), 4. Lynch describes a similar dystopia in *Wasting Away* (prologue).

108. Noteworthy exceptions are few but include Angelo Van Bogart, From the Editor, *OCW*, 4 June 2009, 6; and David E. Davis Jr., "Don't Junk the Clunks," *C&D*, December 2009, 24.

109. The buyers' strike: McCarthy, *Auto Mania*, chap. 7.

110. Lucsko, *Business of Speed*, chap. 9; McDonald and Gosswein interview; Igoe, Letter of the Month (quote).

Coda

1. Bruce Caldwell, "Mustang Market," *HRM*, December 1978, 25–28; "Our Drag Day 5," *HVW*, March 1980, 37–40; Rich Kimball, Collector's Corner, *HVW*, January 1983, 75–76.

2. Caldwell, "Mustang Market"; Rich Kimball, Collector's Corner, *HVW*, October 1984, 16–17; "Ten Best Trivial Pursuits," *C&D*, January 1988, 86–87; Bruce Simurda,

"Semaphore," *HVW*, September 1996, 6; Kevin Oeste, "Killer Paint without Killing Your Budget," *HRM*, April 2002, 24, 26, 28, 30–33.

3. Ken W. Purdy, *The Kings of the Road* (Boston, 1949), 3–4.

4. Eugene Jaderquist, "The Classic Thrill of Yesteryear," *True's Automotive Yearbook*, 1952, 2–7, 100–102 (quote on 100).

5. Tom Igoe, Letter of the Month, *HRM*, September 2004, 16. On the 1970s–1990s, see David N. Lucsko, *The Business of Speed: The Hot Rod Industry in America, 1915–1990* (Baltimore, 2008), chaps. 8–10.

6. Brock Yates, "A Closer Look at Today versus the Good Old Days," *C&D*, December 1977, 22. See also Lee Kelley, "Editorially Speaking," *HRM*, August 1980, 4; and Peter Egan, Side Glances, *R&T*, May 1984, 32.

7. Martin T. Toohey, Inside *Car and Driver*, *C&D*, June 1975, 8.

Essay on Sources

An examination of the role of salvage yards among enthusiasts, *Junkyards, Gearheads, and Rust* is largely based on periodical sources published by and for hot rodders, customizers, import and sports-car fans, old-car hobbyists, and street rodders. For the period after World War II, when the bulk of the developments in chapters 2 through 6 took place, this body of literature is staggeringly vast. Online indexes do exist for a few of these titles, but they were of limited use because the sorts of things that I was looking for often do not show up in tables of contents or indexes: ads for salvage and used-parts firms, unlabeled images of cars in fields or salvage yards, technical articles that discuss cars or parts from wrecking yards, and so forth. The only way to reliably find what I was after was to comb through stack after stack of physical magazines, page by page.

Over the last seven years I therefore spent more time than I care to even attempt to calculate paging through full (or very nearly full) runs of a number of titles, including *Hot Rod Magazine*; *Popular Hot Rodding*; *Car and Driver*; *Road and Track*; *Motor Trend*; *European Car*; *MPH*; *Eurotuner*; *Dune Buggies and Hot VWs*, more commonly known as *Hot VWs*; *VW Trends*; *Street Scene*; and *Cars and Parts*. Some of these titles were only published for a short while—*MPH*, for example, ceased operations after less than three years in 2006. But most have been around for much longer: *Road and Track* since 1947; *Hot Rod* since 1948; *Motor Trend* since 1949; *Cars and Parts* since 1957; *Car and Driver* since 1961 (its predecessor, *Sports Cars Illustrated*, dated back to 1955); *Hot VWs* since 1967; *European Car* since 1991 (predecessors *VW and Porsche* and *Volkswagen Greats* dated back to the 1970s); and so on. I also consulted partial runs of a number of other titles geared toward specific enthusiast subgroups, including hot-rod, custom, and high-performance magazines like *Car Craft*, *Chevy High Performance*, *Popular Customs*, *Hop Up*, *Rod and Custom*, and *Super Chevy*; import and sports-car titles like *Speed Age*, *Motorsport*, *Sports Car Graphic*, the British *Classic and Sports Car*, and *Classic Motorsports*; old-car restoration periodicals like *Old Cars Weekly*, the Canadian *Old Autos*, and the telephone-book-like *Hemmings*

Motor News; street-rod-oriented titles like *Street Rodder* and *Rod Action*; several general-interest titles, including *Automobile, Autoweek, Car Life, Cars,* and *Motor Life*; and the aftermarket journal *SEMA News.*

For chapters 2 through 6, *Cars and Parts, Hot Rod, Hot VWs, Popular Hot Rodding,* and *Street Scene* proved to be the most useful, closely followed by *Car and Driver, European Car,* and *VW Trends.* Chapter 1 makes use of a few of these postwar titles as well. However, since a large portion of that chapter examines the evolution of the wrecking-yard business during the 1900s–1930s, a period long before any of the aforementioned periodicals existed, I turned instead to another set of titles: the general-interest *Horseless Age,* which debuted in 1895; the dealer-oriented *Motor Age,* which began in 1899; *The Automobile,* a general-interest title that appeared in 1899; the Ford-centered *Fordowner,* which became *Ford Owner and Dealer* in 1920, *Ford Dealer and Owner* in 1925, *Ford Dealer and Service Field* in 1926, and *Ford Field* in 1939 as it gradually morphed into a dealer-centered trade periodical; and the industry-oriented journal *Automotive Industries–The Automobile* (later, *Automotive Industries*), the result of a 1917 merger between *The Automobile* and *Horseless Age* (for a brief time during World War II, *Automotive Industries* appeared as *Automotive and Aviation Industries*). *Automotive Industries,* still in publication, proved to be useful at several points in later chapters as well.

Fortunately, the main library at Auburn University has a full run of *Automotive Industries* dating back to the 1940s. To access full runs of *Horseless Age, Motor Age, The Automobile, Automotive Industries–The Automobile,* and the various Ford-centered publications, I began with the online archive at hathitrust.org. This collection, which runs through the end of 1922, features full-page scans in a user-friendly portal that mimics the experience of paging through actual, physical copies. I needed a steady supply of Advil to deal with the searing headaches that came from squinting at my screen for weeks on end, but in research time and money saved, this online portal was a godsend. After exhausting the hathitrust.org collection, I then traveled to the Benson Ford Research Center at The Henry Ford in the summer of 2013, where I was able to read through full runs of *Motor Age, Automotive Industries,* and the several Ford titles from 1923 through the end of the 1930s.

As for the many postwar publications, I already owned several full and partial runs from the research that went into my first book. Rather than travel to a number of libraries to fill in the gaps, as I did when working on *The Business of Speed*, I relied instead on the Auburn University library, which has a full run of *Car and Driver* and partial runs of *Hot Rod*, *Road and Track*, and others—some in print, others on microfilm; miscellaneous copies borrowed from friends and relatives, especially titles like *Motorsport*, *Classic and Sports Car*, and early issues of *Road and Track*; and, most of all, eBay. I bought a number of issues of *Hot Rod*, *Popular Hot Rodding*, *Hemmings*, *Hot VWs*, and others through eBay, as well as more than three decades of *Cars and Parts*. Half of my house now looks and smells like a library, but in terms of the time and money saved, not to mention the convenience of working at a careful pace from home rather than a frantic pace on the road, it's tough to beat buying rather than visiting.

In addition to periodicals like *Cars and Parts*, chapters 5 and 6 also rely on a number of newspaper articles published coast to coast during the 1980s, 1990s, and 2000s. To locate and access them, I methodically searched through ProQuest, LexisNexis, and Articles First using their full-text functions and a long list of key terms designed to generate as many results as possible. This returned thousands of hits in newspapers from Maine to Miami and from New York City to Los Angeles. Careful plodding through these results yielded a number of genuinely useful articles on land-use disputes, clashes over scrappage, and other legal and regulatory matters.

Chapter 6 and, to a lesser extent, chapter 5 also rely on archival and other unpublished resources. Chief among these are the collections at SEMA's Washington, D.C., office, as well as a number of unarchived and unpublished documents at the offices of the California Air Resources Board in Sacramento. At SEMA's headquarters in Diamond Bar, California, I was also able to access a nearly full print run of the industry-insider *SEMA News*, which I then fleshed out using SEMA's online archive of more recent issues at sema.org. Interviews with several key players at SEMA and *Hot Rod* also filled in critical gaps, as did government sources available in print through Auburn University's library system and digitally from arb .ca.gov (CARB's website), www.bts.gov (Bureau of Transportation Statistics), census.gov, www.dot.gov, epa.gov, and gpo.gov.

A number of enthusiast-oriented books on salvage yards, barn finds, and the like proved useful as well, including Pat Kytola and Larry Kytola's *Diamonds in the Rough* (1989), *Cars and Parts' Salvage Yard Treasures of America* (1999), Will Shiers's *Roadside Relics* (2006), Herbert W. Hesselmann and

Halwart Schrader's *Sleeping Beauties* (2007), Michael E. Ware's *Automobiles Lost and Found* (2008), and Tom Cotter's many titles on the barn-find phenomenon: *The Cobra in the Barn* (2005), *The Hemi in the Barn* (2007), *The Vincent in the Barn* (2009), *The Corvette in the Barn* (2010), *The Harley in the Barn* (2012), *50 Shades of Rust* (2014), and *Barn Find Road Trip* (2015). On the question of the old and the new, Ken Purdy's *The Kings of the Road* (1949) and John Keats's *The Insolent Chariots* (1958) were invaluable. So, too, on the question of obsolescence and waste, were Vance Packard's *The Hidden Persuaders* (1957), *The Status Seekers* (1959), and *The Waste Makers* (1960); John A. Kouwenhoven's *The Beer Can by the Highway* (1961); and Peter Blake's *God's Own Junkyard* (1964).

The academic literature on obsolescence, waste, recycling, and the old and the new is vast; the notes to this book's introduction illustrate this point quite well. Rather than list them all again in this space, let me instead note those that were especially useful for this project: Michael Thompson's *Rubbish Theory* (1979); Douglas Harper's *Working Knowledge* (1987); Kevin Lynch's *Wasting Away* (1990); Susan Strasser's *Waste and Want* (1999); Tom McCarthy's "Henry Ford, Industrial Ecologist or Industrial Conservationist?" (2001) and *Auto Mania* (2007); Carl A. Zimring's *Cash for Your Trash* (2005), "Neon, Junk, and Ruined Landscape" (2008), and "The Complex Environmental Legacy of the Automobile Shredder" (2011); the essays in Charles R. Acland's edited volume *Residual Media* (2007); David Edgerton's *The Shock of the Old* (2007); and Kieran Downes's "'Perfect Sound Forever'" (2010).

Rather large as well is the scholarly literature on zoning and land-use conflicts. Here I leaned most heavily on Walter Firey's "Ecological Considerations in Planning for Rurban Fringes" (1946); Robert H. Nelson's *Zoning and Property Rights* (1977); William A. Fischel's *The Economics of Zoning Laws* (1985); John O'Looney's *Economic Development and Environmental Control* (1995); Herbert Inhaber's *Slaying the N.I.M.B.Y. Dragon* (1998); Gregory McAvoy's *Controlling Technocracy* (1999); and Robert J. Johnston and Stephen K. Swallow's edited volume, *Economics and Contemporary Land Use Policy* (2006).

The academic literature on old-car scrappage is also rather large, especially among economists and policy analysts. Especially useful for the present book were Shi-Ling Hsu and Daniel Sperling's "Uncertain Air Quality Impacts of Automobile Retirement Programs" (1994); Robert W. Hahn's

"An Economic Analysis of Scrappage" (1995); Anna Alberini, Winston Harrington, and Virginia McConnell's "Estimating an Emissions Supply Function from Accelerated Vehicle Retirement Programs" (1996); and Jennifer Lynn Dill's "Travel Behavior and Older Vehicles: Implications for Air Quality and Voluntary Accelerated Vehicle Retirement Programs" (2001) and "Estimating Emissions Reductions from Accelerated Vehicle Retirement Programs" (2004).

The scholarly literature on the automobile and automobility is similarly vast. Of particular value to this project were John B. Rae's *The American Automobile* (1965); H. F. Moorhouse's *Driving Ambitions* (1991); Robert C. Post's *High Performance* (1994); John DeWitt's *Cool Cars, High Art* (2002); Kevin L. Borg's *Auto Mechanics* (2007); Steven M. Gelber's *Horse Trading in the Age of Cars* (2008); John A. Jakle and Keith A. Sculle's *Motoring* (2008); and my own *The Business of Speed* (2008).

Index